H. Morford

Morford Short-Trip Guide to America

H. Morford

Morford Short-Trip Guide to America

ISBN/EAN: 9783337145323

Printed in Europe, USA, Canada, Australia, Japan

Cover: Foto ©Andreas Hilbeck / pixelio.de

More available books at **www.hansebooks.com**

MORFORD'S
SHORT-TRIP GUIDE

TO

AMERICA.

(United States and Dominion of Canada.)

BY HENRY MORFORD,

*Author of the "Short-Trip Guide to Europe," "Over-Sea,'
"Paris in '67," &c.*

NEW YORK:
SHELDON & CO., 677 BROADWAY.

LONDON:
W. H. SMITH & SON, 186, STRAND; S. P. BEETON, 300, STRAND.

Entered at Stationers' Hall, London,—all rights reserved.

Entered according to Act of Congress, in the year 1872,

BY HENRY MORFORD,

In the Office of the Librarian of Congress, at Washington, D. C.

James Sutton & Co., Printers, 23 Liberty St., N. Y.

TABLE OF CONTENTS.

COST AND TIME of Short American Trips.............................	7
PREPARATIONS for the Trip to America................................	17
WHAT TO DO and avoid on Shipboard.................................	30
BELL-TIME at Sea	40
NEW YORK CITY, Harbor and Suburbs................................	42
ROUTE NO. 1 —Northern—New York to Niagara and Canada........	62
" No. 2.—Northern—New York to Niagara and Canada........	83
" No. 3.—Northern—New York to Saratoga, Lake George, &c.	91
" No. 4—Eastern—New York to Boston, by New Haven, Providence, &c..	103
" No. 5.—Eastern—Boston to Portland and Canada...........	124
" No. 6.—Northern and Eastern—Boston to White Mountains, &c..	129
" No. 7—Northern and Eastern—New York to White Mountains, &c...	136
" No. 8—Near Western—New York to Philadelphia...........	144
" No. 9.—Western and Southern—Philadelphia to Baltimore, Washington and Richmond.................................	162
" No. 10.—South Western—Richmond to Charleston, Savannah, New Orleans &c...	189
" No. 11.—South Western—Washington or Richmond to Nashville and New Orleans......................................	195
No. 12.—Western—New York or Philadelphia to Cincinnati and Chicago...	197
" No. 13.—Western—New York to Chicago, &c................	206
" No. 14—Western—Cincinnati to Louisville, St. Louis and Chicago...	217
No. 15.—Northern and Western—Buffalo to Chicago, by Lake Shore...	226
No. 16.—Northern—New York or Philadelphia to Coal and Oil Regions...	228
No. 17.—Canadian and Western—Niagara to Detroit and Chicago...	231
" No. 18.—North Western—Chicago to St. Paul and Lake Superior...	237
" No 19.—Canadian—Niagara to Toronto, Ottawa, Montreal, Quebec, &c...	241
No. 20.—Canadian—Niagara to Toronto, Montreal and Quebec, by Steamers..	257
" No. 21—Far Western—Chicago to Omaha, Salt Lake City and San Francisco..	261
OFF ROUTE and Minor Places..	277
DISTANCES, Time and Fares...	304

INTRODUCTORY.

The preparation of the "Short-Trip Guide to America," has been induced by the practical success of the Guide to Europe, on the same plan, and the generally admitted want of some hand-book for tourists in America, fitted for pocket-use, and railway, carriage or steamer reading, by convenient size, clearness of type, and giving the data most ordinarily required, while carefully avoiding tedious and unnecessary details. Such a book has long been required as a necessity, especially for

1st. The very large and constantly-increasing body of English-speaking people, principally from the British Islands, crossing the Atlantic for a short sojourn in the New World, and desirous of seeing the greatest possible variety of interesting places within a limited period, without undue expenditure; and

2nd. The only-less-numerous body of Americans who have made but partial acquaintance with their own country, and who need intelligent guidance in the selection and traversing of the most attractive routes.

The author-proprietor believes that this volume, prepared with much care and labor and after years of very extensive travel on both Continents,—will be found to meet the requirements of both these classes; affording no small amount of information and assistance, meanwhile, to those who visit America for longer sojourn and have no occasion to economize either time or money. It has of course been impossible, in a work of no greater size, to give all the routes of the Continent; but *the most interesting* have been carefully traced, and especially those most likely to meet the views of summer travelers, whether natives or from abroad; while hints have been abundantly supplied, throughout, for other tours, and longer ones, for the benefit of the more leisurely.

To one feature attention is especially called: a paper immediately preceding the Index—"Off Route and Minor Places,"—in which those desirous of visiting, for local or personal reasons, towns or natural curiosities not embraced in the Index or in any of the regular routes, will be likely to find the places required, with brief but sufficient directions for reaching them.

New York City, January, 1872.

THE
Short Trip Guide to America.

The Short-Trip Guide to America.

COST AND TIME OF SHORT AMERICAN TRIPS.

SEVERAL important questions are involved, with Europeans, and especially with Englishmen, in the calculations preceding a trip to America: so that *Whether to go?* precedes the corresponding queries, *How to go?* and *Where to go?* The distance is known to be great, between the Old and New Worlds, though it is really only about one-eighth of that around the globe.

With many men *Time* is the great object, and the want of it the great hindrance; though they may annually spend quite as much of it as would be necessary for a Summer tour across the Atlantic, in lingering about home watering-places and sea-shore resorts, re-visiting the often-seen Lake Country, the Welsh, Irish and Scottish Mountains, etc., or repeating old experiences on the Rhine, among the Swiss Alps or the Pyrenees. This, too, at a time when the great Continent of the West has been made so much more broadly accessible, and so much more closely linked to that of Europe, by the Pacific Railroad, the Atlantic Cable, and other enterprises—when the late great civil war in the United States has necessarily left many fields worth visiting and relics worth gathering — and when steam-transit between the two Continents has become so rapid and reliable that the ocean-passage is little more than that of a ferry. This false idea of *Time* is, as already said, the bugbear which hinders many of those who have comparatively liberal means and a fair proportion of leisure; but with a far greater number of those who love Nature in her varying moods and wide differences, and who desire to see the different peoples of the world, *at home,—Money* is the anxiety, the want of it the hindrance, and the belief that a mint is necessary for anything in the shape of transatlantic travel, the great bugbear which confines them to one continent.

A large proportion of this is a mistake, originally induced by want of intelligent inquiry, and materially added to by the exaggerations, not to call them falsehoods, of some of those who have been over the desired routes. While "going to America" was principally confined to the wealthy few or those driven by business demands, it was at once an easy and a tempting thing to do, to add to the supposed importance of what had been done, by overstating the cost as well as enlarging on the personal adventure and peril; and, truth to say, the habit has not yet quite died out, now, when the many follow in the track of the few and detection is so much easier. Many a man, of quite the average integrity, but who supplies (as he believes) the center at home of an admiring circle, not many members of which are likely to follow him abroad—cannot resist the temptation to show, when he returns, that he has been doing, in the way of cost, what *they* had better not attempt if they do not wish to fail miserably; and it is just possible that there have been members of the opposite sex, guilty of adding to the misunderstanding by corresponding exaggerations of their own elevation above the untraveled and, consequently, the easily-deceived.

Travelers tell "travelers' stories," in a pecuniary as well as an adventurous point of view: that is the truth, briefly stated; and those stories frighten away many who would else enlarge their knowledge of life by seeing other continents than their own.

Now it is the fact that the European can spend much money in America, within a very brief period and without going over any wonderful space, if he will; just as in travels on the Continent, years ago, the average Englishman spent twice as much, under the same circumstances, as the man of any other nation, creating in different minds the impression of his being a "prince" and a "fool"—until the American became first his rival and then his admitted superior, in the detail of lavish and tasteless expenditure, and the Englishman who made any pretensions to common sense, taking a lesson from the example, comparatively abandoned the field of extravagance. The lavish and the reckless may still pave their very way with gold, if they will—as evidenced by the fact that a certain well-known Englishman, spending less than three months in the United States, drew upon his London bankers during his absence, for nearly £3,000: the expenditure all the more notable, because the tourist, a markedly free liver and entertainer, made no purchases of consequence for preservation, did not play, and never indulged in what are called the "costly vices." Others have followed, in different approximations, ranging between £300 and £800 the month of absence; though it is to the credit of the national wisdom, to say that these instances of what must be considered wasteful expenditure for any one not in possession of a princely patrimony or a great banking-house, are somewhat rare.

So much for what may be spent in very brief tours, by those who can afford plenty of money, or think that they can do so: now for what may be saved, or rather for the question upon how little these brief tours may really be made, without discomfort or painful compromise of position.

Even in the steerage, on some of the best-appointed lines, passages may be made with much less discomfort than most stay-at-home people suppose; and it is not at all certain that thousands of hardy persons, limited in means, who spend the requisite amounts of time and money on very questionable home-amusements, approaching to vices, might not do well to tempt a little rugged life in the forward parts of the ships that carry over their wealthy brothers in the saloon-cabins. For on the best lines the discomforts, inconveniences and unhealthiness of steerage-passage have all been materially ameliorated within the past three or four years: the sleeping accommodations on many of them are endurable if no more; the food is almost always plentiful and generally excellent; the amount of amusement enjoyed is always greater than that attainable by the better-lodged people at the stern; and the safety to person is necessarily the same except under circumstances of gross carelessness.

Let us see, for the benefit of those very limited in means and still desirous to see a little fragment of the New World—what would be the absolute cost of doing what emigrants of both sexes and all countries very often do for the sake of spending a few days with friends in the places of nativity. Say that six weeks' time is attainable, and let the cost of that six weeks be measured as carefully and yet as liberally as possible.

Steerage passage to New York, £6 10s.—return, £6 10s.; total, £13. Time not on board ship, about three weeks; board, for that time, average of £1 15s. per week, £5 5s. Expenses of sight-seeing about New York, Boston and Philadelphia, with conveyance to each, during that period, £5. Occasional necessary conveyance, the feet being principally trusted to, £3. Incidental expenses, liberally calculated, £4. Total. £32. £10 to £12 more would enable the cheap tourist to visit Niagara Falls, greatest of American natural curiosities, and see at least one or two cities of the Queen's dominions in Canada. Grand total, with that included, £42 to £45, with a certainty that any economical person, in good health and temper, could come within the smaller sum named, and even reduce it, without other inconvenience than carefully adhering to the cheaper rates of conveyance corresponding to that of the ocean transit.

How many comparatively-poor men are there with longing and hopeless desires after seeing other countries than their own, who never make any calculation or effort to such an end, and yet who could and would compass it if they fairly understood the comparative trifle for which so much might be enjoyed!

One of the greatest of American travelers, Mr. Bayard Taylor, made his first European excursion under circumstances quite as illiberal as anything here indicated—"did" Great Britain and a very considerable portion of the Continent on foot, except with rare instances of riding, and remained not less than six or seven months, his whole expenditure being only about $500 (say £105), and the fortunate result of his travel that successful volume "Views a-foot; or, Europe Seen with Knapsack and Staff." And it is very doubtful whether in any portion of his later experience, in all descriptions of traveling "state," up to that of Secretary of Legation at St. Petersburg, he has ever enjoyed his wanderings better than when making that first essay as a poor boy. As a pleasant pendant to which, the writer recalls having met, not many months ago, in the streets of New York, an English workman from Sheffield, spending a little of his moderate surplus-earnings in seeing what he called "a tidy bit of the New World," and one of the most intelligent of travelers on many subjects of interest and enquiry —whose expenditure, as given by himself, would not reach within ten per cent. of the figures above given, while he was healthy, happy and entirely comfortable in what his saving expenditure allowed him to secure and enjoy.

But the figures already given represent, of course, the minimum possibility of travel in any desirable part of America, compatible with even the decencies of life, without too many of its comforts; and, it is, equally of course, with that class of people standing midway between the possible steerage-passenger and the traveler *en prince*, that we have next and principally to do. The most important question of this paper is—*What need be the expenses and the time consumed for a certain round, of a traveler going first-class and demanding all the comforts, and yet indisposed to waste money on costly luxuries?*

To answer that question, then, as intelligibly as may be consistent with brevity.

For six to seven weeks' absence from home, visiting New York, Boston, Philadelphia, Baltimore, Washington, one or two of the most noted watering-places, Niagara Falls, and one or two of the cities of Canada—only

Ticket to New York and return, £36 to £60—say an average of £48, for which all necessary comfort and quite sufficient "style" can be secured. Average board of the three weeks off-ship, £3 to £4 per week —say £10 10s. Traveling expenses, railway and carriage fares, etc. £30. Maps, pictures, curiosities, etc., (not at all necessary, but inevitable), £10. Incidental expenses, for which no name can be given; money to guides, beggars, stewards and servants; money lost and wasted, with an occasional indulgence in a luxury, *not* including costly wines or "society," gambling or other vices—£15. Total, £113 10s. £12 to £15 or possibly £18, may easily be saved from this, by a *very* careful person, leaving the expenditure about £100; and a person at all the reverse of careful may quite as easily *add* a corresponding sum, making the expenditure, with no greater amount of travel or sight-seeing, £125 to £130.

For ten weeks' absence, an estimate of £50 additional may safely be made, bringing the total outlay up to say £160 10s; and this will secure, in addition to the round already named, an extension of the tour through the White and Green Mountains of the north-east, with Saratoga, Lake George and Lake Winnepisaukie; or it will add the great Coal Regions of Pennsylvania, Cincinnati, and other cities of Ohio, with Chicago and glimpses of the Mississippi and the Great Lakes.

For three months' (thirteen or fourteen weeks)

absence, another £50 may be added, bringing the amount up to say £200 or £210; and with this all the foregoing may be done, with the addition of the "North-West," now found in the States bordering the Upper Missouri, with the Lake Superior region, and a much more extended visit to the cities of Canada, and the natural curiosities of the Dominion; or, it will enable the tourist (if the season should be a proper one for Southern travel) to go southward from Washington to Richmond, Charleston, Savannah and New Orleans, with their intermediate towns and a general view of what is technically known in the United States as "the South."

Four months will add to this £50 to £60 of expenditure, bringing up the outlay to £260 or £280, and permitting the pursuance of some of the routes named, more at leisure, as well as the addition of others of the watering-places, if the visit is paid in the proper season. And within the same time may even be managed a run over the Pacific Railroad, to Salt Lake City, San Francisco, and the great natural curiosities of California, with a view of the Pacific —though five months would be a more rational calculation for the whole time of absence. With the California route added, the expenditure will be found materially increased from all the previous calculations—say £100 additional for that alone; the amount necessary for the four-to-five-months trip, with the Pacific excursion crowning it, being somewhere within the range from £360 to £380 or £400, and half of the continent travelled over in that time and at that cost.

At this point the phrase "short-trip" may be said to be exhausted; for only people of liberal means and abundant leisure are likely to go far beyond in any one visit, and to them these calculations possess only limited interest; though even they may find a certain advantage in bestowing that slight amount of study on the subject, necessary to secure a proper knowledge of time to be spent and money used to the best purpose.

PREPARATIONS FOR THE TRIP TO AMERICA.

The following paper, like some of the others to come after it, is especially intended for those who have never before crossed the Atlantic, and, consequently, some of the advice tendered in it may seem very primitive to those who have already taken their degree, however low a one, in the academy of traveling experience. The suggestion may properly be added, however, that even some of those who have taken that degree may find themselves none the worse for reading over these hints, even if they do so to dissent from them. An apology may need to be made, too, for the direct and conversational style adopted in this and some other papers: the aim of the writer is, in this regard, to come as near as possible to the words and manner that would be used in a personal conversation, with one of the parties doing much more than half of the talking.

It may be proper, too, with reference to this paper and those succeeding, to say that the writer speaks almost entirely from personal experience—and that where that experience has failed, it has been eked out, not often through the means of books, but from the personal hints and relations of frequent and experienced travelers. For himself, the writer, in repeated and extended travel on both continents, has made, first and last, nearly all the mistakes against which in the present instance he attempts to guard others, and felt the necessity for some instruction like that which he now endeavors to impart, on almost every point touched upon. So much said, the promise of the paper must be kept, in a brief but comprehensive list of rules connected with the preparation for transatlantic voyages, and especially for those *first* voyages which more or less imitate Columbus.

1st. Decide whether you can afford time and money to go at all, taking into consideration the before-urged opportunities for economy. Also, decide whether, in going, you leave too much of anxiety, personal or pecuniary, for fair enjoyment; for there is an old adage about the absentee who "drags with each remove a lengthening chain," and there are not charms enough, even in the natural scenery and odd character-studies of the New World, to make such a trip "pay," when the heart or the business-powers must be left at home. So much decided, and in the affirmative, then

2nd. Having made up your mind, stick to the resolution. Arrange your time of going and make everything work to accommodate *that*, not leave that to accommodate itself to everything. Generally, in this as in everything else in life, too long anticipation is not the healthiest or the most profitable, and a voyage not canvassed over for five years in advance is likely to yield more pleasure than one submitted to that length of speculation. Above all things never boast that you are going, when you have merely *thought of going* and made no definite decision; as unpleasant consequences may often result, in the event of the projected voyage being abandoned, and the suspicion may sometimes creep into the minds of acquaintances, either that there was "bounce" in the original statement, or that some heart-failure at the last moment has induced the abandonment.

3d. Having resolved upon time of going and probable duration of trip, and selected the line of steamers by which the outward voyage is to be made, do not permit the paltry folly of wishing to keep a certain number of pounds for a few days longer in pocket, to prevent the early taking of a passage. The best state-rooms of any favorite steamship are likely to be first taken up; and even in the event of any unexpected hindrance occurring, there is rarely any difficulty in disposing of a well-located berth, while most of the companies, at any time before the "eleventh hour," will transfer the passenger from one steamer to a later one, if a change of time is all the deviation from the original plan rendered necessary.

4th. If suddenly-occurring events happen to have changed the calculation in the other direction, and the plan of going is formed almost at the very moment when some favorite steamer is about to sail, never heed the stories so likely to be told, that "the steamer is full and you cannot get a place!" There is nearly always room for "one passenger more," as there is in most land-conveyances; and if the worst comes to the worst, it is a very rare case when some of the officers of the ship cannot be found ready to give up a room for the run, at the inducement of no-very-large addition to the price of the passage-ticket. These are suggestions for extreme cases, however: as before said, passage had much better be taken early, whenever possible, for reasons already stated or about to be.

5th. In selecting berths, when a good opportunity for choice remains, always aim to get as near as possible to the midships of the vessel—a consideration of not much consequence to old voyagers with strong nerves, but of great importance to landsmen, as every foot of distance from the waist increases the amount of motion in a heavy sea; and not only is the danger of sea-sickness less amidships, but the chances of having sleep broken by the "pitch" of a "head" or "following" sea are proportionably decreased when so located. The same principle applies, in a less degree, to the question of outside or inside rooms (those inside or outside of the alleyways). There is much less effect from the "roll," in a "beam" sea, for those occupying inner berths; but there is always much less light for reading or any other purpose, and the one advantage will probably balance the other, except in winter passages, when the inner rooms are altogether preferable.

6th. No guide-book, until very lately, ever contained a hint of the advice to be embodied in this paragraph; and yet there is no word of counsel, of the whole, more important. Unless that miserable being, a "man of letters," and thus compelled to be always reading—there are few intending voyagers, male or female, who will not be the better for a little "reading up" as to any country about to be visited. A fresh glance at the atlas, to see how the different sections lie and the relation which they bear to each other, is almost indispensable, even to some of us who flatter ourselves (before we think the second time) that we learned our geography in early life, and have kept pretty well up with it ever since. This rule, as already hinted, applies to travel and to travelers in all countries but to no other geographical division of the earth's surface with the same force as to the New World, and especially to that covered by the great Republic of the West, where change seems the rule, and where the alterations of boundaries and names, within the past few years, have been almost as startling as those effected in political status and society. Upon the relative positions of different States, the names of their capitals and chief cities, their rivers and natural wonders and even their commercial and industrial features, it is exceedingly profitable to be as well as possible freshened in advance; and the same remark obviously applies with equal force to the main facts of history and the more important points in established or current literature. And to the latter suggestion a few words of particulars may be added. Exactly as a man from the New World would find more than half his possible pleasure lacking, visiting Great Britain without possessing any comprehensive knowledge of Shakspeare, Scott, Burns, Thackeray, Moore, Tennyson, etc., or France with no acquaintance with Rousseau, Voltaire, Lamartine, Beranger, Victor Hugo, Dumas, etc.—so the visitor from the Old World to the New must be lacking in many of the opportunities for observation, social life and popularity, who fails to know something of Cooper, Irving, Hawthorne, Longfellow, Whittier, and (especially for travel in the New England States) Holmes, Emerson and Lowell.

7th. Another "rubbing up" is advisable though not absolutely indispensable. Thousands of questions about native land, its physical appearance, wealth, working of government, industrial aspects, etc., are constantly asked of all persons on their travels, supposed to be of the average intelligence, by foreigners whom they chance to meet; and it is decidedly pleasant as well as proper, not to be three or four thousand miles from home, unable to answer the simplest questions with reference to things occurring at our own doors. The more we know about our own land, the more intelligent and agreeable travelers we shall make; and, in this connection,

8th. Throw overboard two false impressions, together, before leaving Europe. Overboard with the idea, at once, that the land you are leaving is better than all others in every regard, so that nothing can be learned abroad: and with it give the go-by to the alternative impression that you have nothing worth asserting and even boasting about, and that what you are to learn abroad will stand in place of the previous experience and pride of a life. Each of the leading European countries possesses, at this day, many things unequaled by the rest of the world and matters of legitimate pride to her citizens; but she is almost equally sure to have errors and deficiencies which may well be corrected by observations among other, if not necessarily wiser, people. Every tourist going abroad should carry with him all practical knowledge of his own land, and all well-founded pride in it; and, at the same time, he should travel with eyes and ears open and power to divest himself of ridiculous national vain-glory prejudicially shutting away all beyond.

And now to a few minor particulars belonging to the very eve of starting, and still important enough to deserve place and number:

9th. Start with a confident expectation of returning, and yet leave property-interests disposed of as if no return was likely to be made. There is really less danger, in a given number of days, in going over-sea than in most railway-travel; but absences thus involved are necessarily much longer and demand additional forethought in at least one or two particulars. "No man dies the sooner for making his will," they say; and certainly no man travels less comfortably for leaving affairs at home in such a shape, that, if he does not return, his absence will cause the least possible inconvenience to those left behind. And, in this connection, again,

10th. There is nothing wiser for the departing "family-man," whatever the status of those dependent upon him, than an investment in a moderate *life-assurance*, with an additional *assurance against accident*. Nothing of an earthly character (the religious questions will naturally suggest themselves) adds more comfort in a storm at sea, or danger in some distant land, than the reflection that there would be, at least, one benefit from the risk terminating unfavorably: *the dear ones at home would be pecuniarily the gainers by it.*

11th. Arrange baggage compactly, and not too extensively. For each person (male—the ladies *will* make rules for themselves, applying what hints may chance to suit them)—one stout leather or wood-and-leather trunk of 30 to 36 inches by 16 to 20 inches, and one convenient valise for carrying in the hand, are always sufficient, for anything less than carrying over the whole personal effects with a view to residence. The trunk for deposit in the great cities, in the event of expecting to return along the same line—if not, unavoidably to be carried along. The valise for short excursions from those great centers having this advantage—that it can be carried in the railway-carriage or cab, while the trunk must be looked after, with trouble and expense. Both trunk and valise should be plainly marked with name and residence—initials not always enough for either safety or convenience. If the trunk is small enough for the sea-voyage to find place in the state room, all the better; if not, care must be taken that, before it goes into the hold, all articles are taken out from it that will be needed before landing on the other side. The valise will always find place in the state-room, of course. And this brings

12th. The important question of Clothing, with reference to which a few general suggestions may be found valuable. The point of view here taken is especially for the male sex, but the female will find it easily varied to their requirements. For crossing the North Atlantic, to return in two or three months, the first requirement is a suit of thick clothes, so old and valueless that one can lounge upon the deck in them, with no fear of damage. (Dandyism is at a discount at sea—a lesson quickly and surely learned). Clothing thick, because sea air is nearly always damp, and generally cold. Then as thick an overcoat and gloves as can well be procured, the use of which will become patent, either off the Irish coast or among the fogs and possible icebergs of the Banks of Newfoundland. A thick blanket, rug, or heavy robe, to make lounging upon deck the easier and warmer. For summer travel in the Northern, Middle and Western States, or Canada, a neat traveling-suit of Melton, with one of flannel for proceeding further southward, and for very hot weather in even the States named. A summer-overcoat or wrap of waterproof Melton or *aqua-scutum*—not so regularly or often needed as in the British Islands, but indispensable. Heavy-wool under-clothing for sea-use, with courage enough to double it if necessary; for American hot weather, on land, lighter under-clothing of merino, silk, or zephyr. A dress-suit, if entrance into "society" is intended, or if there is plenty of room in the trunks; as clothing is somewhat high in America, while exceptionally tasteful and well made—though, candidly, in hurried trips of this character, the traveling-suit is seldom shaken off. Figured or colored-wool overshirts, with high throat, collar and wrist-bands, for time at sea, or for any temporary "camping-out" or "roughing-it" among woods or mountains. Plenty of linen and white goods, to avoid being at the mercy of the washerwoman at times of sudden transit, and because all these, as well as all hosiery and under-clothing, cost more in the United States than in England. Stout-soled shoes—of calf, best. Low-crowned tourist-hat, of felt (dress-hat to be bought, if necessary); umbrella, of late years almost as indispensable on the western side of the Atlantic as the eastern; a good opera-glass, necessary for catching views rapidly and correctly, both by sea and land, and more convenient if not too large for the pocket and not necessitating the *prononcé* strap.

13th. Make such arrangements, if possible, that a little longer absence than that contemplated will not work serious business or other inconvenience, as the best calculator cannot always be quite sure of non-detention through some influence or action beyond himself.

14th. Arrange (as before suggested) to take a little more money abroad than is supposed to be necessary for either time or distance; but

15th. Carry in actual money, (English gold, or Bank of England notes, with a trifle of silver) only so much as will pay expenses on ship-board and last during the few days that may happen to elapse before reaching the point at which the first draft is made payable. All beyond this should be taken either in bills-of-exchange on bankers in one or more of the more important cities to be visited, or in circular letters-of-credit to corresponding houses in those cities. It is scarcely necessary to say that only the very first class of banking-houses, at home, should be dealt with, in procuring exchange or letters-of-credit, if the painful possibility of finding oneself abroad without funds, is to be avoided.

16th. Take some letters of introduction, when tendered, and to the *right persons;* but depend very little upon them, except in some business point of view. The fact is that, without any discourtesy to givers being intended, letters of introduction go for less in America and secure less consideration additional to the deportment and standing of the bearer, than in any other part of the civilized globe: and they should be understood and rated accordingly. Added to which may be set down that in no other country is the best society of any given region so accessible, the letter of introduction being thus rendered little else than commercial or useless.

17th. Avoid attempting to carry over, among baggage, anything that can be construed as beyond necessary personal use, as the American administration of the customs, of late years, is stringent to oppressiveness, and a misunderstanding on that point may be more easily avoided than removed. (Articles most watched for and guarded against are clothing, [new and in undue quantity,] silks, linens, laces, watches, jewelry and precious stones).

18th. Create as little impression as possible, on the verge of departure, of feeling that some event, moving half the world, is taking place in your first leaving your native land. A sea-voyage, now, no further than America, is about equivalent to a trip from London to Edinburgh or Dundee, fifteen years ago—and not much more than was the transit across the channel to France, at the distance back of thirty or fifty years; and the observing world is generally coming to regard it in that light.

19th. and last. If possible, go on board before the last moment of sailing, and have any heavy luggage on board even earlier. Also, if possible, make any extended tender farewells earlier and elsewhere than on the crowded deck of a steamer, at the last moment, when everybody is in the way of everybody else, when the officers naturally wish to throw overboard all the whiners, and when there is a probability of the grief of departure being added to by the worry of having wife, sister, child or friend tumbled into the river at the landing-stage or dropped over between tender and steamer as the two separate.

WHAT TO DO AND AVOID ON SHIPBOARD.

The advice in this paper, too, will be set down didactically, and much of it will be considered as very elementary by those who have once or oftener crossed the Atlantic. In the meantime, not even to some of them will the maxims be found unprofitable, if attended to—judging by the very large number of habitual travelers who seem to happen upon the very conditions of discomfort and imprudence, as if seeking them.

1st. Perhaps the first condition of comfort in a sea-voyage, is to avoid making up the mind as to any positive time at which the voyage must be concluded. To look across the three thousand miles of the Atlantic, and think over the days necessary to travel it, even on the swiftest vessel, is rather discouraging than the reverse, to people of rapid thought and active habit; but by simply avoiding any definite calculation and considering the ship and her officers and crew as doing their "day's work," the amount of impatience may be very considerably reduced. Creeping ahead a little every day, the whole voyage will soon be accomplished: that is enough to know and enough to feel, no matter what anxieties may be at the end.

2d. Perhaps the next desideratum is to avoid any

considerable anxiety as to the voyage being a prosperous one, by first remembering that more than an hundred runs are made without a single accident, and more than five hundred without the total loss of a vessel—and then falling back upon that pleasant recollection that you have not the affair in charge, any way—that (Providence over all, and always to be remembered, of course,) the officers and crew of the ship have their duty to do and are very likely to do it, for the sake of their own lives and the property committed to their skill. It may be straining a point, perhaps, but there is really some philosophy in getting into the state of mind of the droll fellow who demonstrated to one of the "anxious," in a storm at sea, that, having paid their passage, and the company having consequently contracted to take them across, the question of the ship's foundering was really something with which they (the passengers) had nothing to do! This may not have much reassured the frightened man, but it certainly silenced him; and there no doubt was more than a grain of earnest in the old traveler's philosophy of *remembering that he did not steer the ship*, as there was undoubtedly comfortable indolence in it.

3d. It is wise not to expect too much on shipboard, either in the way of luxury, or even of positive comfort. Ships, at the largest, are small as compared with hotels, and at the steadiest are "shaky," as compared to private dwellings, except when the latter have the rarity of earthquakes to throw them off the perpendicular. Plenty of good food, respectable though confined sleeping-quarters, and attendance fair, but by no means that of a first class hotel —these are all that ought to be expected; and a very little philosophy makes them enough. It has before been said that "dandyism is at a discount, at sea;" so is, or ought to be, *finickiness*. What if neither shaving, nor dressing, nor any of the other offices of civilized life, can be done quite as well as at home? Nobody notices whether they are scrupulously performed, or not; and some of the neatest of men when on shore, when they have become old travelers, consent to be slovenly for those few days without serious suffering. The golden rule, on going to sea, is: *Expect very little, and be prepared to bear good-humoredly with it;* then, if "all the modern conveniences" should happen to prevent themselves, as is not at all likely, they will afford double enjoyment, and the want of them will not entail misery.

4th. Determine to be as jolly as health will allow, and as companionable as is at all consistent with the temperament. Join in all practical harmless amusements and exercises, with the result of making your own days less tedious, and producing the same effect on those of others. One jolly fellow, sometimes, seems to leaven up a whole ship-load; one or two glum faces act like a wet-blanket on all concerned. There is a comradery in sea-going, scarcely second to that of the army; and some of the pleasantest friendships of years originate on the deck filled with comparative strangers. Quoits, shovelboard, chess, draughts, backgammon, social games at cards, all these supply amusement to those who will take part in them; and there is room for any amount of table sociability at meals, not marred but rather increased by the little accidents to which breakfasting or dining in rough weather is certainly subject.

5th. Make friends, early, with the captain and other officers of the ship, so far as they will permit; but take no liberties with them, and carefully avoid compromising any one of them who may have shown any peculiar favor, by speaking of it to others of the ship's company or passengers. Strictly observe those cardinal rules which forbid going upon the bridge, talking with the officers when on duty, or distracting the attention of the quarter-masters at the wheel. Avoid getting in the way of the officers at the compass, or hindering them when engaged in that most important event of the day—"taking an observation." Obey them, quietly and respectfully, when they give a direction calculated to secure your safety or prevent accident—even if the reason of the order should not be fully evident to a landsman. Don't inquire any oftener than is unavoidable, where the ship is at any particular moment, what a certain movement on deck means, what kind of weather it is going to be during the next twenty-four hours;

and don't ask the men, when they are heaving the log, how many miles an hour the ship is going, or don't expect them to tell the truth if you do! Don't get in the way when hawsers are being overhauled or yards braced; and don't wonder if, getting in the way when some evolution of hauling the ropes is going on, you occasionally trip and so learn what times and places are dangerous. Don't attempt to "help," at any time, except in the rare event of an accident; and thus "keeping out of the way," without losing any chance of observation and enjoyment, secure the friendship of the officers, the respect of the crew, and the gratitude of all concerned.

6th. Make friends with the stewards, at once, not only by treating them respectfully, but by speaking to the two in charge of your particular table and state-room—requesting their attention and promising them the due *douceur* at the end of the voyage. Half a sovereign each to the saloon and lower-saloon stewards, and say a crown to the "boots," with half-a-crown for beer to the captain of the watch who first "chalks" you when you break the rules of the ship by going forward, and perhaps half a dozen shillings to persons who do errands for you during the run—this, reaching not much more than thirty shillings altogether, is quite sufficient to grease the wheels of service and make welcome then and afterwards.

7th. Avoid attempting to read much, at sea, however interest may tempt in that direction. There is a motion and jar of the vessel, making the letters swim and damaging head and optic-nerves to a degree needing days for recovery. Some persons can read steadily, almost without injury: others cannot: it is never best to try the experiment when it can be avoided. And there is rarely much occasion: it is a poor passage-list in which more amusement cannot be found than in books, for the short period consumed in crossing the Atlantic.

8th. Keep on deck, all that is possible. Half the charm of going to sea lies in the pure, fresh air, except in very stormy weather. The air of lower-cabins and state-rooms is necessarily more or less confined, and consequently unhealthy; while the healthiest atmosphere in the world comes fresh to the lungs from blue water. There is far less danger of sea-sickness, too, on deck than below, when actual illness does not enforce confinement to the berth; and the thousand sights and sounds of sea-life—sunrises, sunsets, moonlight, storm-waves, whales, porpoise-shoals, passing vessels, observations, log and lead heaving, making and taking in sail, signalling, etc., are only to be enjoyed by those who keep the deck as persistently as possible. And this is even additionally true at times of leaving or making land; approaching port, etc., opportunities for remark and study, lost during which periods, may be and probably will be lost forever.

9th. Dress warmly—quite as warmly as comfort demands, and err on the safe side if at all. Sea-air, though healthy, is damp and deceptive as to temperature. Never mind the appearance: put on the clothes.

10th. Take much exercise. Want of occupation induces long sitting at table and hearty eating; and the system must be a strong one which can endure this for days, without exercise, and yet suffer no injury. When there is not too much sea to make it possible, at least a mile or two should be walked every morning and a corresponding space in the afternoon—the long cleared decks, or the alleyways, of most of the best steamers, rendering this *amusement of exercise* easy and convenient.

11th. Aid the direction last named, by eating moderately as usual habits will allow—either by abridging the quantity of each meal, or by avoiding some of the number. Four meals per day are usually provided—breakfast, lunch, dinner and tea: very often, and especially when there is any tendency to inactivity of the system, and fever, two of the four may be profitably omitted.

12th. Put confidence in the ship: believe, for the time being, that the ship is the best afloat. If you go down into the fire-room (which, by the way, is quite as well kept out of), don't fall into the fancy that so large a mass of fire in the midst of a vessel must inevitably burn her: vessels are especially constructed to guard against that danger, and iron does not take fire easily. Don't be alarmed at the noises continually coming from the fire-room, or think that some calamity has happened there: firemen are normally noisy as well as grimy, and they need to speak loudly to make themselves heard. Don't fancy, in short, that everything will go wrong unless you attend to it, except in one particular; and that is,

13th. Join the fire-police of the ship, and stick to the organization. Take no combustible materials below in your baggage—neither matches or dangerous chemicals; take no light of any kind below the decks, for better reason than because there is a severe punishment for any proceeding of the kind— the all-powerful reason that such an act may destroy your own life and the lives of others. On this point, watch your own conduct and that of others, and no harm is likely to result from the close surveillance; though any discovery made should always be communicated quietly to some person in charge, and not shouted through the ship so as to create a panic among the passengers.

14th. Never go forward when the ship is pitching into a heavy sea: there is always danger of injury, in such an experiment by a landsman, and very often of being swept overboard, at times when even sailors can scarcely keep footing on the wet and slippery decks. Never stand at or very near the taffrail (extreme stern) in correspondingly heavy weather, as there is always danger of the ship "jumping out from under you"—an accident which sometimes happens to experienced seamen who stand unguardedly in that dangerous position. Never climb upon the bulwarks, however calm the sea; for there is no knowing at what moment there may be *one* roll—enough to finish the individual voyage very unpleasantly.

15th. Never attempt to go up or down one of the companion-ways (stairs), or along one of the gangways, or the decks, when the sea is heavy, without making as much use of the hands as the feet—holding on firmly to the nearest convenient rail. Broken ribs or limbs are sometimes the consequence of forgetfulness or bravado, on this point.

16th. In the event of illness (other than sea-sickness), don't take nostrums, or trust to anything in your private "medicine-chest." There is always one surgeon, or more, on each ship; they are paid for attending to the health of passengers, without charge except for costly medicine; they are particularly familiar with the treatment prudent at sea; and it is very often the case that medicines upon which dependence can be placed when on the more stable element, prove injurious in the abnormal condition of never being entirely quiet.

17th. If sea-sick, don't fancy the disease is a mortal one. Few people die of it, though many (it is to be feared) are rendered vastly uncomfortable. Keep the bravest heart and the strongest determination possible, against the great foe; and above all, do not join the noble army of those who ask to be mercifully "thrown overboard" as a means of escaping the torture. Nobody dares obey the request—not even your worst enemy, who wishes that he could; and if it *should* be obeyed, the chances are ten to one that before you had gone down ten fathoms in blue water the cry might be a different one.

18th. Berths, in sea-going ships, are mostly single; and yet it is best, especially in heavy weather, to have a *bed-fellow*. This is easily found in the valise or well-filled carpet bag, which packed closely in against the side-board, the would-be sleeper lying on the side in the inner part of the berth, will generally enable him to lie without rolling, even when the ship is doing her worst in that direction, and secure sleep when it would be otherwise impossible from the constantly-waking motion. An alternative arrangement of almost equal excellence in rough weather, though not always practicable—is to use a broad luggage-strap, fastened to any stanchion at the back of the berth and then buckled around the breast of the would-be sleeper.

19th—and more important than any of the preceding. Remember, oftener than when the service is read on Sabbath morning, that there is a Hand, wiser and stronger than that of any officer of the ship, ruling not only the vessel, but the waves upon which she rides and the winds and other elements which may place her in peril.

BELL-TIME AT SEA.

PASSENGERS by any of the transatlantic steamers, or on any other extended route involving the continual change of longitude, should never risk injuring their time pieces by setting them slower or faster, but quietly allow them to run down immediately after starting, and keep them in that condition, though carrying them in the ordinary upright position, until the end of the voyage. They will be obliged, meanwhile, to depend upon the ship's bell, with occasional glimpses of the saloon-clock, for the requisite knowledge of the flight of time during each day, to prevent a mental vacuum on that subject, and enable them to make proper preparation for meals.

A little experience of the use of the bell, however, is necessary for putting this advice into ready practice; and the following brief table of "bell-time at sea" will be found worth an hour or two of study, to that end; one fact being always borne in mind: that the farther eastward the faster the time, and, the farther westward, the slower; so that a steamer of ordinary speed loses about half an hour per day of the running time with which she is charged, in going eastward, and gains a corresponding amount of time in going westward.

Commencing the day at sea, with the half-hour succeeding midnight, the following explanation of the "bells" (*i. e., strokes of the bell*) will be found easily understood and quite sufficient for practical use, if one aid to the memory is employed—the recollection that the odd numbers of strokes are always half-hours, that the even numbers are always hours, and that those hours which can be divided by 4 are always represented by numbers which can also be divided by 4.

1 bell	....	½	o'clock,	A.M.	1 bell	....	½	o'clock, P.M.
2 bells	....	1	"	"	2 bells	....	1	" "
3 "	....	1½	"	"	3 "	....	1½	" "
4 "	....	2	"	"	4 "	....	2	" "
5 "	....	2½	"	"	5 "	....	2½	" "
6 "	....	3	"	"	6 "	....	3	" "
7 "	....	3½	"	"	7 "	....	3½	" "
8 "	....	4	"	"	8 "	....	4	" "
1 bell	....	4½	"	"	1 "	....	4½	" "
2 bells	....	5	"	"	2 "	....	5	" "
3 "	....	5½	"	"	3 "	....	5½	" "
4 "	....	6	"	"	4 "	....	6	" "
5 "	....	6½	"	"	1* bell	....	6½	" "
6 "	....	7	"	"	2 bells	....	7	" "
7 "	....	7½	"	"	3 "	....	7½	" "
8 "	....	8	"	"	4 "	....	8	" "
1 bell	....	8½	"	"	1 bell	....	8½	" "
2 bells	....	9	"	"	2 bells	....	9	" "
3 "	....	9½	"	"	3 "	....	9½	" "
4 "	....	10	"	"	4 "	....	10	" "
5 "	....	10½	"	"	5 "	....	10½	" "
6 "	....	11	"	"	6 "	....	11	" "
7 "	....	11½	"	"	7 "	....	11½	" "
8 "	....	12	noon.		8 "	....	12	midnight.

* From 4 P. M. to 8 P. M. instead of presenting an unbroken succession of bells from 1 to 8, is divided into two "Dog Watches"—4 to 6 ("first dog-watch") and 6 to 8 ("second dog-watch")—in order to prevent the larboard and starboard watches of sailors being on duty during the same hours, one day after another—as they would be if they were continually and only changed once every four hours.

NEW YORK CITY, HARBOR AND SUBURBS.

APPROACH AND HARBOR.

LAND is generally made, approaching the harbor of New York, from any vessel coming down the "Great Circle," at some point on the Long Island coast, at starboard or right of the ship; and the time may be anywhere from four to ten hours (in clear weather) before crossing the bar at Sandy Hook, the entrance of the Lower Bay of New York. After first sighting, this land will keep in sight—low and uninteresting, the course of the vessel being nearly parallel with the shore, and at a few miles distance. Pilots are taken on board from small schooners, at distances varying from a few miles from the coast to two or even three hundred—as disasters from want of pilotage off this port, many years ago, have induced much activity and competition, of late years.

Two to three hours from Sandy Hook, for ships coming down the Long Island coast, and as a first sight for those crossing from the south, are made the *Highlands of Navesink*, fine bold headlands approaching the sea, and forming one point of the eastern coast of New Jersey. These hills show to excellent advantage on a nearer approach, and are very imposing when the Bar at Sandy Hook is being crossed, two square-tower lighthouses showing on the Highlands, behind the long, low point of wooded sand forming the Hook, on which are to be seen one light-house and two beacons, with a formidable line of Government fortifications in progress, near the outer or northern end, very near to which the ship necessarily passes the channel.

Passing the Bar and running up the Lower Bay, the New Jersey Highlands continue ahead and to the left, sloping away towards Long Branch a few miles southward ; on the right continues *Long Island*, with the still lower and sandier *Coney Island* adjoining it in front ; still ahead and to the left rise the hills of *Staten Island*, with an opening between it and Long Island marking the *Narrows*, through which entrance is made from the Lower or Raritan Bay into the Upper or Bay of New York proper.

At the left, four or five miles below the Narrows, is passed (if there is no occasion to make its nearer acquaintance) the *New York Quarantine*—a range of low buildings on an artificial island built within the last few years on a shoal known as the West Bank of Romer. Passing the Narrows, the fine fortification to the right, on Long Island, is *Fort Hamilton*, with the ruins of the once celebrated *Fort Lafayette* standing in the water at some distance below it—while to the left rises the corresponding bluff of Staten Island, crowned with a light-house and fortifications, with a strong new structure, *Fort Richmond*, standing below at near the water's edge.

The view of *New York Bay*, after passing the Narrows, is considered one of the finest of its character in the world, and should never be lost by the traveler enjoying the opportunity for the first time. On the right, passing up, will be observed the Long Island shore, handsomely shaded, and dotted with the residences of well-to-do citizens or suburbans; and on the left Staten Island presents much higher ground, landings and thriving villages near the shore, and the sides of the hills in like manner well shaded and dotted with tasteful residences. Some six miles above the Narrows, at the immediate right, the monuments of *Greenwood Cemetery* may be seen covering and crowning one of the Long Island hills near the shore; still to the right, but ahead, the *City of Brooklyn* shows its many spires and wilderness of buildings; immediately ahead rises *Governor's Island*, with its round fort, *Castle William*, and its long ranges of barracks and officers'-quarters; and as Governor's Island is passed, still directly ahead, the *City of New York* is seen, stretching right and left, from its lowest point at the Battery, up the East and North Rivers (Long Island Sound and the Hudson), each line showing a perfect forest of the masts of shipping, and the marked deficiency of commanding spires partially relieved by the nearness and grace of that of Trinity Church.

From this point, which best reveals the splendor of New York Harbor, Brooklyn lies a little behind, at the right; Staten Island has fallen away to a much greater distance behind and at the left; the Hudson River stretches northward, immediately ahead, Long Island Sound branching away eastward at an acute angle; the other two islands of the harbor, so far unnamed, *Bedloe's* and *Ellis'*, lie at some distance to the left; and behind them, to the left and ahead, on the west or New Jersey side of the river, may be prominently seen the towns of *Jersey City* and *Hoboken*, continual high lands rising up-river from the latter, along the Hudson, towards Fort Lee and the Palisades.

It is also at this point that the traveler visiting the New World for the first time from the Old, will find one of the most marked of sensations in observing not only the immense variety of shipping and the flags of all nations at the wharves and in the stream, but the many particulars in which the American river and ferry craft differ from those of any other nation—the prevailing color being white, and both strength and grace often sacrificed to speed and temporary convenience.

NEW YORK CITY AND BROOKLYN.

As will already have been observed, the City of New York lies at the junction of the North or Hudson River and Long Island Sound (familiarly called the East River), having thus the best of opportunities for cleanliness and health, which are by no means always embraced with due diligence and faithfulness—the city being always ineffectually cleaned, in comparison with the cost to the people, and often disgracefully dirty. In effect, Brooklyn, immediately opposite on the southeast, and connected with it by half-a-dozen or more well-managed steam-ferries, is a part of the same city, though lying in another county, and bearing a different name; while nearly the same may be said of both Jersey City and Hoboken, on the New Jersey shore, and reached in the same manner by ferry.

Before proceeding to explore the city or suburbs, it should be noted that carriage-service in New York is very high and very bad; cab-service better and improving, though by no means up to the European standard—so that the first should be almost entirely avoided, and the latter much oftener foregone in favor of the public conveyances than they would be in any city of the Old World. The ferries should be used freely, not only for necessary crossings, but as an additional means of studying the topography of the harbor, and the excellence of the system. For most directions the street horse-cars run regularly and well, and are comfortable, except at morning and evening hours, bringing too great crowds; and, on Broadway, the omnibuses are available and respectable.

Of Streets, the best worth noting is *Broadway*, which should be driven, in open carriage if convenient, from its commencement at the Battery (harbor-side) to its virtual termination at Union Square, many of the best commercial buildings being thus seen. Thence *Fifth Avenue* should be taken, to the Central Park, a view being thus caught of the finest fashionable street in America, and one of the handsomest in the world, though very irregular in architecture. Much of the leading fashion of the city may be found gathered in the streets running out from Fifth Avenue, from Fourteenth to Sixtieth streets—notably on *Twenty-third*, *Forty-second* and other wide streets. *The Bowery* may be noted as the people's or east-side Broadway. *Greenwich street* will be found filling a somewhat similar position on the west side; *Third, Sixth* and *Eighth Avenues* may be taken as fair types of prosperous commonalty and bustle; *West street* (Hudson River side) will be found to supply a jam quite worthy of the Strand at its worst hours; and still further down town, *Wall street, Broad* and *New streets* command attention as the centers of the moneyed interest. In BROOKLYN, the most notable streets are *Montague* and *Clinton*, for fashion; *Fulton, Court* and *Atlantic streets, Myrtle Avenue,* &c., for business activity; *Third street, Union street, Fourth Avenue,* &c., as drives; *Clinton, Washington, Bedford, Grand* and other *Avenues,* for suburban beauty.

Of Wharves, New York has none that are not thoroughly contemptible, though there is promise of this default being gradually remedied, under new arrangements employing the talent of General McClellan and other engineers. Of Markets, few that are not disgraces as to erection and keeping—the best exception being *Tompkins Market*, at Third Avenue and Seventh street; though none in the world have more variety as to supply, than *Washington Market*, foot of Vesey street, Hudson River side, and *Fulton Market*, foot of Fulton street, on the East side. Of Museums, none except that at Central Park, and the small but unique collection at Brooklyn Navy Yard. Of Libraries—the *Astor*, an inconvenient and overrated humbug; the *Mercantile*, for merchants; the *Society;* and one or two minor ones of little consequence. Of Public Galleries, none but the somewhat extensive ones of the picture-dealers, *Schaus*, Broadway; *Knoedler*, and *Somerville*, Fifth Avenue, &c., except during annual exhibitions of the *Academy of Design;* though some arrangements are in progress for a permanent free gallery, of merit and importance, and the Private Galleries of Messrs. A. T. Stewart, John Taylor Johnson, W. T. Blodgett, Aspinwall and others, are very creditable and sometimes exhibited to the public. Large collections of national and celebrity portraits are to be seen in the great photograph galleries of *Brady* and *Fredericks*, Broadway; *Gurney*, Fifth Avenue, &c. Of Hospitals, only the inconvenient *Bellevue*, at Twenty-sixth street and East River, since the cruel demolition of the *New York*, Broadway and Duane street; *St. Luke's* (a comparatively private benevolence); *St. Vincent's*, and one or two minor ones of little consequence.

Of Educational Institutions and the structures connected, the following are most notable: *Columbia College*, (an institution of moderate age but reputation and usefulness, and with Law and Mining Schools attached) East Forty-ninth street; *New York University* (collegiate, but making no pretence to fill the European use of that word), Washington Square; *New York College* (formerly the New York Free Academy), Twenty-third street and Lexington Avenue; *College of Physicians and Surgeons*, East Twenty-third street and Fourth Avenue; *University Medical College*, Worth street; *Rutgers Female College*, Fifth Avenue; *Union Theological Seminary*, University Place; *New York Law Institute*, Chambers street; *Protestant Episcopal Theologi-*

cal *Seminary,* West Twentieth street ; &c., &c. In connection with educational facilities it should be added, that the Common Schools of the City of New York are the best in the world, free to all, numerously attended, and worth observation by any visitor.

Of Monuments, New York has as follows : In Central Park (hereafter mentioned) *Humboldt, Schiller,* &c. In Union Square, equestrian statue of *Washington,* by Browne, and statue of *Lincoln.* In Madison Square, monument obelisk to *Gen. Worth.* In Trinity Churchyard, *Martyrs' Memorial* (handsome Gothic structure in honor of revolutionary patriots who died on the prisonships); monument to *Captain Lawrence,* who fell on the Chesapeake ; and horizontal slab over the remains of the heroine of the romance of the same name, *Charlotte Temple.* In St. Paul's Churchyard, shaft to *Robert Emmett,* the Irish patriot ; monument to *Gen. Montgomery;* one (back of church) to *George Frederick Cooke,* the actor. In Printing House Square, bronze statue of *Franklin,* presented to the Printers of New York by Capt. Albert De Groot.

Of Antiquities, the city may be said to have literally none, the hand of "improvement" having lately been very busy with the few remaining. The two most interesting old buildings existing, are the *Old Walton House,* Pearl street, most fashionable residence of the past century, now decayed ; and the *Washington Hotel,* Broadway and Battery Place, once the residence of Gen. Washington, of Sir Guy Carleton, &c.

Of Churches, few command any attention architecturally, though there is no deficiency as to number. The two oldest are the *North Dutch*, Fulton and William streets, now about being demolished, and the *Middle Dutch*, used as a prison by the British during the War of the Revolution, and now the city Post Office—Nassau, Liberty and Cedar streets. *St. Paul's*, Broadway (where the pew of General Washington, when President, still remains), and *St. John's*, Varick street, best deserve present notice, from age and unpretending grace ; and *Trinity*, Broadway, as the most respectable finished Gothic erection on the Continent—though *St. Patrick's Cathedral*, Fifth Avenue and Fiftieth street, will eventually dwarf it and all others. Those remaining, best repaying visits of curiosity, are *St. George's*, Rutherford Place ; *Grace Church*, Broadway ; *St. Paul's*, and *All Souls*, Fourth Avenue ; *St. Thomas'*, Fifth Avenue ; *Holy Trinity*, Madison Avenue ; *St. Mark's* (old) Stuyvesant street ; the *Tabernacle*, Sixth Avenue ; *St. Stephen's*, Twenty-eighth street ; *Dr. Chapin's*, Fifth Avenue. In BROOKLYN (named, from their numbers, the "City of Churches)", the most notable are the *Holy Trinity* and *St. Ann's-on-the-Heights*, both on Clinton street ; *Dr Eddy's*, Pierrepont street ; *Church of the Pilgrims*, Henry street.

Of Public Buildings the most interesting, from one cause or another, will be found the *City Hall*, City Hall Park (with a collection of civic and heroic portraits of some interest, in the "Governor's Room"); the *New Court House* (unfinished, but with many handsome rooms) same place ; the *City Prison* ("Tombs"), Centre street;

the *Custom House* and *Sub-Treasury*, Wall street ; the *Cooper Institute*, junction of Third and Fourth Avenues ; the *Bible House*, opposite the preceding, above ; the *Academy of Music*, Fourteenth street ; the *Academy of Design* and *Christian Association* buildings, Fourth Avenue and Twenty-third street ; *Booth's Theater*, Twenty-third street ; the *Grand Opera House*, Eighth Avenue ; *Tammany Hall*, Fourteenth street ; the *Central Police Station*, Mulberry street ; *Hudson River Railroad Freight Depot*, Hudson street (with colossal bronze of much oddity and a singular merit, on the principal front, in honor of Cornelius Vanderbilt) ; New *Grand Central Depot* of the Harlem, Hudson River and New Haven Railroads, Fourth Avenue and Forty-second street ; new *Post Office* (building), lower end of City Hall Park ; *Methodist Book Concern*, Broadway and Eleventh street ; *Masonic Hall*, (building), Twenty-third street and Sixth Avenue ; *Stock Exchange* (new) Broad street ; *Produce Exchange*, Whitehall street. In BROOKLYN, the *City Hall* and *County Court House*, Court and Fulton streets ; *Academy of Music*, Montague street ; *Mercantile Library*, and *Academy of Design* (new) same street ; *Atheneum* Atlantic street, etc.

New York has many Commercial Buildings of great cost and splendor—no other city in the world having more of what may be designated as "palaces," devoted to money or trade. The lead is taken among purely financial buildings, by the *Park Bank*, Broadway. No less than three structures devoted to Life Assurance command much attention—those of the *Equitable Society*, at Broadway and Cedar street ; of the *Mutual Company*,

Broadway and Liberty street ; and of the *New York Company*, Broadway and Leonard street ; while several other Banks and Insurance Buildings, recently erected, on Broadway, Wall street, Nassau street, William street, Pine street, &c., deserve only less attention. The most prominent among what are known as the "business palaces," are those of *A. T. Stewart*, Broadway and Chambers street, and Broadway and Ninth street ; of *Lord & Taylor*, Broadway and Grand street, and Broadway and Twentieth street ; of *Arnold & Constable*, Broadway and Nineteenth street ; of *Tiffany*, Union Square and Fifteenth street; of *Ball & Black*, Broadway and Prince street ; of the *Waltham Watch Company*, Bond street ; of *Appletons*, Broadway ; of *Brooks Brothers*, (old "Maison Dorée") Union Square, &c.

There are many Private Dwellings of great cost, splendor, and varying architectural taste, on Fifth Avenue and the more fashionable streets on Murray Hill ; the first among them being the recently completed palace of *Mr. A. T. Stewart*, at Fifth Avenue and Thirty-fourth street, of which the details, without and within, are of the most lavish magnificence, while the picture collection embraces Church's "Niagara," Rosa Bonheur's "Horse Fair," Yvon's "America," Dubufe's "Prodigal Son," &c. Those of *Mr. George Opdyke*, Fifth Avenue and Forty-seventh street ; *Mr. William M. Tweed*, Fifth Avenue and Forty-third street ; *Mr. Wm. H. Vanderbilt*, Fifth Avenue and Fortieth street ; *Messrs. Phelps, Dodge, and Phelps*, Madison Avenue, Thirty-sixth and Thirty-seventh streets ; *Sig. Barreda*, Madison Avenue and Twenty-fifth street—all deserve attention for costly elegance. Of Club Houses, the three most prominent are the *Union*, Fifth Avenue ; the *Union League*, Madison Avenue ; and the *Manhattan*, Fifth Avenue.

Of Hotel Buildings, (also Hotels) New York has many of great size and fine architecture ; prominent among them being the *Gilsey House*, Broadway and Twenty-ninth street ; the *Fifth Avenue*, Broadway and Twenty-fourth street ; the *Grand Hotel*, Broadway and Thirtieth street ; the *St. Cloud*, Broadway and Forty-second street ; the *Metropolitan*, Broadway and Prince street ; the *Sturtevant*, Broadway and Twenty-eight street ; the *St. James*, Broadway and Twenty-sixth street ; the *Westmoreland*, Union Place ; the *Coleman*, Broadway and Twenty-seventh street ; the *Westminster*, Irving Place ; the *Grand Central*, Broadway opposite Bond street ; the *Everett House*, Union Square ; the *St. Nicholas*, Broadway and Spring street ; the *Astor House*, Broadway and Vesey street ; the *Brevoort*, Fifth Avenue ; the *New York*, Broadway and Fourth street ; *Western* and *Merchants'* (both mercantile) Cortlandt street ; &c. In BROOKLYN, the *Pierrepont House*, Montague street, and the *Mansion House*, Hicks street. Two Newspaper Offices of mark are to be noticed—that of the *Herald*, at Broadway and Ann street ; and that of the *Times*, at Printing House Square.

The principal Theaters of New York City proper are *Wallack's*, Broadway and Thirteenth street ; the *Olympic*, Broadway near Bleecker street ; *Niblo's*, Broadway near Prince street ; *Booth's*, Twenty-third street and

Sixth Avenue ; the *Grand Opera House*, Eighth Avenue and Twenty-third street ; the *Fifth Avenue*, Twenty-fourth street ; *Wood's Museum*, Broadway and Thirtieth street ; *Union Square Theater*, Union Square ; and the *Bowery*, and *Stadt Theater*, Bowery. Opera Houses, the *Academy of Music*, Fourteenth street, and occasionally the *Grand Opera House*. Ethiopian Minstrel House, *Bryant's Opera House*, Twenty-third street. BROOKLYN has several excellent places of amusement, in the *Academy of Music*, Montague street ; *Brooklyn Theater*, Washington street ; *Park Theater*, Fulton street, *Hooley's Opera House*, Court street, etc.

Most popular Churches (for service) *Trinity*, Broadway (Episcopalian); *Grace*, Broadway and Tenth street (Episcopalian); the *Tabernacle*, Sixth Avenue and Thirty-fourth street (Cong.) ; *Dr. Chapin's*, Fifth Avenue and Forty-fifth street (Univ.); *St. Thomas'*, Fifth Avenue (Epis.); *Dr. Tyng's*, Rutherford Place (Epis.) ; *Fifth Avenue*, Fifth Avenue and Nineteenth street (Pres.); *St. Paul's*, Fourth Avenue and Twenty second street (Methodist Epis.); and in BROOKLYN, *Plymouth* (Rev. Henry Ward Beecher's) Orange street (Cong.) ; *St Ann's-on-the-Heights* and the *Holy Trinity* (Epis.) both on Clinton street ; *First Baptist*, Nassau street ; *Dr. Talmadge's Tabernacle*, Schermerhorn street. Present Catholic Cathedral, in New York, *St. Patrick's*, Mulberry and Houston streets ; with other leading Catholic Churches, *St. Stephen's*, Twenty-eighth street near Third Avenue (noted for fine music), and *St. Francis Xavier's*, Sixteenth street, near Fifth Avenue.

Public Grounds—*Central Park*, (see "Excursions," following); *Washington*, *Madison* and *Union Squares*, and *Battery* and *City Hall Parks*, most of limited dimensions, but all assuming attractive shapes, and most of them being provided with music on certain evenings of the week, during the warm season; and *Jones' Wood*, lying on the eastern side of the Island, on the river, opposite the lower end of the Central Park, with fine woods in and around, and famous as a place for great out-door gatherings, including the German and Irish festivals and the Scottish annual games; *Jerome Park*, Westchester (also see "Excursions"); and in BROOKLYN, *Prospect Park*, (also see "Excursions") *Prospect Park Fair Grounds, Lefferts Park*, &c.

Principal Bazaars (for purchases): *Stewart's*, Broadway and Tenth street (Dry Goods); *McCreery's*, Broadway and Eleventh street (Dry Goods); *Lord & Taylor's*, Broadway and Twentieth street (Dry Goods); *Arnold & Constable's*, Broadway and Nineteenth street (Dry Goods); *Tiffany's*, Union Square (Jewelry, Bronzes, Plate and Works of Art); *Ball, Black & Co.'s*, Broadway and Prince street (Jewelry, Bronzes, Plate and Works of Art); *Stevens'*, Union Square (Jewelry, Bronzes, Plate and Works of Art); *Macy's*, Sixth Avenue and Fourteenth street (Fancy and General); *Lyle's*, Eighth Avenue, and Bowery (Fancy and General).

Other objects of interest to those making longer sojourn: the *East River Bridge*, now building between New York and Brooklyn, and promising to be one of the master-works of its class in the world; the *Pneumatic*

Tunnel, commencement of subterranean travel in the city, to be seen at Broadway and Warren street; *Governor's Island*, head-quarters of the military department; the *Navy Yard*, Brooklyn; the Penal and Charitable Institutions on *Blackwell's, Randall's* and *Ward's Islands* (under control of Commissioners of Charities and Correction—building, Third Avenue and Eleventh street); and a variety of *Asylums* for Orphans and the afflicted.

SUBURBS, DRIVES AND EXCURSIONS.

Of Drives and Short Excursions, (by carriage) the first favorite is that to the

CENTRAL PARK, a large and admirable public ground, occupying nearly the centre of the Island, extending in width from Fifth to Eighth Avenues, and in length from Fifty-ninth to One Hundred and Tenth street, handsomely laid out, shaded and ornamented, with fine roads and costly bridges, and lacking only age to be equal to any public ground in Europe. It has a *Lake*, with boats (service); a *Museum*, with Zoological collection and many other curiosities; a *Casino*, on the European plan; *Public Carriages*, making the round of the Park at short intervals, for trifling fare; *Statues* of Humboldt and Schiller and of Professor S. F. B. Morse, the telegraph promoter; statuary groups of the "Hunter and his Dog," "Auld Lang Syne;" and presents the feature of music by a fine band every Saturday afternoon during the warm season, attracting immense concourses of people. In connection are also to be seen the *Croton Receiving Reservoirs*, alleged to be of size enough, and to contain water enough, to float the navies of the world. [*Central Park* may also be reached from the City Hall, by horse-cars on the Belt railways (along either river), on the Third and Madison Avenues, Broadway, Sixth, Seventh and Eighth Avenues.] Beyond, the drive by carriage is often and profitably extended to the *Harlem* and *Bloomingdale Roads*, or to HIGH BRIDGE, an aqueduct bridge over the Harlem river, of great height and solidity ; or to *Jerome Park*, new and handsome trotting and racing ground of the American Jockey Club, beyond the Harlem River, in Westchester. Another scarcely less fashionable drive is to

PROSPECT PARK, the new but very handsome public ground of Brooklyn, which bids fair to rival if not to excel the Central, has a *Lake*, a *Dairy Cottage* and *Barn*, a fine stretch of natural forest, an elevated drive with commanding view, statues of President Lincoln (at entrance), Washington Irving, &c. Music by a fine band, Saturday afternoons. [May also be reached from New York by Fulton ferry and by horse-cars of Flatbush Avenue line.] Near Prospect Park is to be visited

Greenwood Cemetery, one of the largest and handsomest Cities of the Dead on the globe, with lakes, rising grounds, fine shades, costly monuments, and all the other melancholy attractions possible to be flung around places of burial. Among the leading features are the handsome sculptured Entrance Way ; the Firemen's, Pilots', Old Sea Captain's, Canda, Scribner, J. G. Bennett and McDonald Clarke monuments ; the tombs of William E. Burton, the comedian, Lola Montez (with inscription, "Eliza Gilbert") Crawford Livingston, &c.;

the vaults of Stephen Whitney, William Niblo, &c. Most beautiful point, that at and around "Sylvan Water;" finest views, those from "Ocean Hill" and "Battle Hill." [May also be reached from New York by Fulton ferry and horse-cars of the Greenwood or Fifth Avenue lines.] Beyond Greenwood and Prospect Park, the same drive may be profitably extended to *Prospect Park Race Course*, and to

Coney Island, fine sea-beach, with excellent bathing and somewhat miscellaneous attendance, and not too eclectic in its general character. [May also be reached from New York by Fulton ferry, and by Smith and Jay street horse-cars, or by either of the horse-car lines to Greenwood, thence by steam to the beach.]

Other Short Excursions will be those to the *Brooklyn Navy Yard*, with extensive Dry Dock, Museum of marine curiosities, and much of general interest; to *Fort Hamilton*, at the Narrows, junction of the Upper and Lower Bays, with fortifications and very fine sea-view [drive, or may be reached by Fulton or Hamilton ferry, and horse-cars]; to *Evergreen Cemetery*, East New York [drive, or Fulton ferry and Fulton Avenue horse-cars]; to *Canarsie* and *Rockaway Beach* [Fulton or South ferry, Fulton or Atlantic Avenue horse-cars to East New York, steam cars to Canarsie, and steamboat to Rockaway]; to *Hoboken*, great base-ball and cricket grounds, and favorite German resort, across the Hudson, in New Jersey [ferry from Barclay street]; to *Bergen Point*, [drive or horse-car from Jersey City, or train on New Jersey Central Railroad, from foot Liberty street]; to *Paterson* and *Passaic Falls* [train on Erie Railway, foot Chambers or Twenty-third street : see route North by Erie Railway]; to *Newark*, largest and most thriving city in New Jersey [train on New Jersey Railroad, foot of Cortland street ; or Newark and New York Railroad, foot Liberty street : see route New York to Philadelphia]; to *Elizabeth*, New Jersey [train on New Jersey railroad, foot Cortland street, or New Jersey Central, foot Liberty street]; to *Staten Island*. (*New Brighton* and other popular resorts) [ferries from Battery and from foot Dey street]; &c.

Longer Excursions of interest, conveniently made from New York, those to (1)

LONG BRANCH, great sea-shore resort on the New Jersey coast, with several miles of fine bluff, bold surf-bathing, admired sea-view, splendid drives and excursions, and an immense number of summer hotels, capable of accommodating fifteen to twenty thousand visitors (among the principal the *Continental*, *West End*, *Mansion House*, *United States*, *Metropolitan*, *Howland*, *Pavilion*, &c.), and a present popularity making it the most generally sought and notable place on the American sea coast. It supplies the summer residence of President Grant, and has many cottages of the wealthy. Near it are *Eatontown* (with *Monmouth Park Race Ground* in the immediate neighborhood), *Red Bank*, *Deal*, and other villages of New Jersey. [Reached by boats of the New Jersey Southern Railroad, to *Sandy Hook* (with government fortifications and entrance to the Lower Bay); thence rail, by the *Highlands of Navesink* (fine elevation, with splendid sea-air and view, and summer-boarding place of merit and popularity—hotels, *Thompson's, Schenck's,*) *Seabright*, &c., [the whole distance within sight of the sea.] [From Long Branch railway connection to *Freehold*, and thence to Trenton and other cities of West New Jersey; or train may be taken for *Manchester*, *Tom's River*, and towns of New Jersey further southward; to *Atlantic City* or PHILADELPHIA.] To (2)

LAKE MAHOPAC, pleasant and very popular minor watering-place, with handsome quiet wooded scenery, islands, fine boating, sailing, fishing and other attractions. Reached by Harlem Railroad, in a few hours, through the very fine scenery of that line, at the lower edge of the Hudson Highlands. Hotels, *Gregory House, Baldwin House*, &c. To (3)

SCHOOLEY'S MOUNTAIN, (*Heath House*) mineral springs and popular summer-resort, with fine air and charming scenery, in the minor mountains of New Jersey; reached by the Morris and Essex Railroad, from foot of Barclay street, by *Morristown*, one of the handsomest towns and most popular residences in the State, to *Hackettstown*, whence short ride by stage-coach. Also, *Budd's Lake*, within a few miles of the preceding, and reached by same conveyances—with many attractions of boating, fishing, &c. Also, *Lake Hopatcong*, with similar attractions to the place last named, reached by the same railroad to Stanhope or Dover, thence carriage or boat to destination. To (4)

DELAWARE WATER GAP, (*Kittatinny House*), lying, as the name indicates, at one of the finest passes of the Upper Delaware, through and among the mountains dividing Pennsylvania and New Jersey, and with superb mountain and river scenery, pure and healthful air, and much popularity as a place of summer resort. Has many features of especial woodland beauty, in *Rebecca's Well*, *Venus' Bath* and *Eureka Falls*, views from *Prospect Rock*, *Fox Hill*, &c. [From the Water Gap, continuing by rail, may be reached *Stroudsburg* and the Lackawanna Coal Regions of Pennsylvania; or, southward, *Easton*, PHILADELPHIA, &c.] To (5)

Greenport and *Orient*, minor watering-places at the east end of Long Island; and to *Jamaica* and other places nearer. [Reached by Long Island Railroad.] Also, to *Glen Cove*, and other near places on that Island, by boat. To (6)

WEST POINT, by evening or morning boat or Hudson River railroad. (See route to West Point, Catskills, Albany, &c., Route No. 1.)

ROUTE NO. 1.—NORTHERN.

NEW YORK TO NIAGARA FALLS AND CANADA, BY HUDSON RIVER, NEW YORK CENTRAL RAILWAY AND CONNECTIONS.

Division A.

NEW YORK TO AND AT WEST POINT AND HUDSON HIGHLANDS.

The transit from New York to West Point and the Highlands may be made in from two to four hours, by (1) Hudson River Railroad to Garrison's, then ferry to West Point; or by (2) morning boat on the river, to West Point direct; or (3) evening boat on the river, also direct. Either of the latter is preferable to the former, for reasons hereafter to be given.

By Rail.

Leaving New York by rail, on Hudson River Railroad, the first object of special interest, except the high lands at and about Fort Washington, studded with fine residences,—is the crossing from New York island to the mainland of Westchester, at Kingsbridge or *Spuytenduyvel;* and on the opposite or western side of the river, commence, at about the same point,

THE PALISADES, immense almost perpendicular masses of rock, rising sheer from the river on that side, in shape suggesting the name, and continuing at various heights of hundreds of feet, for some ten miles, where they break away into rugged hills.

Beyond Spuytenduyvel, the first place of importance passed through is the handsome small town of *Yonkers;* then *Dobbs Ferry*, with the long wharf of the Erie Railway opposite, at *Piermont*, and a ferry between; then *Tarrytown* (where the laying over of a train may be well compensated in visiting "Sunnyside," the late residence of Washington Irving, the Major Andre Monument, &c., in the immediate neighborhood); then *Sing-Sing*, with its strong State-prison buildings, and on the opposite bank of the river a view of the gorge running back to the celebrated *Rockland Lake*, from which so much of the best ice is derived. After leaving Sing-Sing, very soon is crossed the *Croton River*, from works on which and the lake of the same name, the New York supply of water is derived. Shortly after crossing the Croton, a mass of rocks, rising conically and crowned with a light-house, on the other or west side of the river, marks *Stony Point*, celebrated for the reckless courage displayed in its capture by Gen. Wayne, during the Revolutionary War. The next stopping-place of importance is *Peekskill*, on leaving which the

HIGHLANDS OF THE HUDSON are entered, presenting their heavy and picturesque masses on both sides of the river, and enchanting the eye with the continual changes, appearances and disappearances made inevitable by the course of the railway through them. The disembarkation from the railway is made at *Garrison's Landing*, whence ferry-boat and omnibus to the Military Academy or one of the hotels at West Point.

By Steamboat.

Precisely the same features as those indicated by rail, will be enjoyed by boat, with the advantage of both sides of the river being seen in lieu of one, and the additional escaping of the noise inevitable in riding by rail along rocky passes. When entering the Highlands, however, the advantage of the boat is even more manifest, as there is scarcely a river or lake approach in the world, so magnificent as that through the Highlands proper, from Peekskill to West Point—fine as any one point of the Rhine, and forcibly reminding the tourist of the middle and upper portions of Loch Lomond, approaching and above Inversnaid. It is from boat on the river, especially, that the alternating wild beauty and rugged grandeur of the giants of the range, their feet at the very water's-edge, can best be appreciated.

Morning boats, making this voyage, and then going on up the Hudson to Albany, leave New York every morning, at an early hour; and evening boats, passing through the Highlands before nightfall, leave every afternoon.

Disembarkation, from either, is made at Cozzens'

or the Military Academy docks at West Point; thence to the hotels by omnibus.

At and near West Point.

One of the principal attractions at West Point, consists in the admirable views which can be enjoyed either from *Cozzens'*, the fashionable hotel and summer resort, on the high cliffs below the Military Academy, the *Parry House*, in the same vicinity, or the *West Point*, above it, making quiet lounging a continued luxury. This is not true of one direction alone, but of all, the elevation being high and the reaches of the river, above and below, singularly beautiful. Of excursions, the most notable is to

Old Fort Putnam, ruins, with some portions of solid wall remaining, lying on a hill westward from the Academy. This fort must always retain its interest, as the "Key of the Highlands" during the Revolutionary War, and the scene of Arnold's intended treason. The views from it, in all directions, too, are the very finest to be enjoyed in any portion of the Highlands. An early visit will, of course, be paid to the

United States Military Academy, which gives the place its peculiar importance, and which ranks among the first of military institutions, with some features of severity attracting peculiar attention. [Information as to modes and forms of visiting, can always be obtained at the leading hotels.] In connection with the Academy comes the interesting spectacle,

Parade of the Cadets (morning and evening)—which should not be missed—the evening especially, by any who desire to see the perpendicular in carriage, the angular in motion, and the sharp in discipline.

Pleasant excursions may also be made to *Buttermilk Falls*, in the neighborhood; and across the river to *Cold Spring*, and to the *Robinson House*, standing four or five miles south from it, where Arnold resided at the time of his treason. Near Cold Spring may also be seen *Undercliff*, residence of the late Gen. Geo. P. Morris, the poet.

Division B.

WEST POINT TO AND AT THE CATSKILL MOUNTAINS.

Northward from West Point, by steamboat on the way towards Albany, from the wharf; or rail from Garrison's Station, opposite. Assuming that the boat will be taken, and remembering that if proceeding by rail the variation of scene will be very slight—the following will be the most important features, beyond West Point. Emerging from the Highlands proper, and passing "Cronest" and "Storm King," the largest hills of the range, and also *Cornwall Landing* on the left, with much beauty and picturesque scenery in the neighborhood (among other attractions, *Idlewild*, residence of the late N. P. Willis), and *Fishkill Landing* on the right, is shortly reached, on the left,

NEWBURGH, very slopingly situated on the high bank, with large river-trade, an important railway connection westward to the Erie road, and one feature of great importance on the bluff below: *Washington's Head-Quarters*, a revolutionary relic of prominence, with many reminders of the hero and the struggle still preserved. Above Newburgh, though the river is fine, there is no feature of marked interest, until, at the right, is reached

POUGHKEEPSIE, a large town with some picturesqueness of location, and a triple distinction compounded of its heavy river-trade in agricultural products, the manufacture of ale, and the proximity of the noted *Vassar Female College*.

Within a few miles after leaving Poughkeepsie, the rough scenery is supplemented and completed by the breaking into view, far ahead and to the left, of the

Catskill Mountain Range, which thenceforth scarcely leaves the eye of the tourist until arrival —so graceful is the outline, and so beautifully blue the general aspect. Minor landings of *Hyde Park*, etc., are passed, to

Rhinebeck, on the right, where landing is made for *Rondout* and *Kingston*, on the opposite side (connection by ferry), and for

The Overlook Mountain House, new but very popular place of summer resort, at great height on the southern portion of the Catskills, and commanding a most magnificent view, especially eastward and southward. Also with many attractive features in the neighborhood, in the *Devil's Kitchen, Cleft in the Rocks, Pulpit Rock, Overlook Cliff, &c.* Also, at a little distance, *Shoe Lake*, a beautiful and attractive sheet of water. [Reached from Rhinebeck by ferry to *Rondout*, thence by rail to *West Hurley;* thence by stage-coach, by the Sawkill Creek and Woodstock, to destination. May also be reached by evening boat, direct from New York to Rondout, thence as before.]

Beyond Rhinebeck are passed *Barrytown* and other landings on the right, *Malden* and others on the left, to

Catskill Landing, point of disembarkation for the Mountains, and of crossing from *Oak Hill Station*, for those who have come up by the rail. Also, popular summer resort, at the *Prospect Park House*, immediately above, with fine grounds and admirable view; at the *Powell House* (posting-house for the mountains, on the wharf) &c. [Catskill Landing may also be reached by evening boat from New York direct, and direct connection made for the mountains.]

From Catskill Landing by stage-coach, always in waiting for boats and trains, by *Catskill Village*, the *Half-Way House*, and at one-third distance of the ascent of the Mountains proper, the *Rip Van-Winkle House*, with a broad flat rock beside it, on which tradition alleges the sleep of Irving's hero to have taken place. Views over the Hudson Valley are very fine, before reaching the

Catskill Mountain House, among the highest of all American places of eastern sojourn, and in many regards the superior of all others on the continent, as to situation. The view from the house, over the Hudson river and valley, is wonderfully extensive and beautiful; and *Sunrise*, as seen from the piazza, is scarcely second to the same spectacle from the famous Swiss Rhigi. Of excursions, there are many and most pleasing. The most interesting (longer ones by carriage, always in readiness) follow. To

Kauterskill Falls, wild and romantic basin, with two cascades, of 180 and 80 feet, and picturesque in every aspect, above and below, besides being surrounded by wild and grand mountain and ravine scenery, and views of *High Peak* and *Round Top*, the two giants of the range, obtainable from different points. (The *Laurel House*, a popular place of sojourn, standing at near the verge of the falls, affords residence to the many who wish to study the splendid scenery in this immediate neighborhood). To the *Lakes*, small sheets of water, lying in primeval wildness; short walk from the Mountain House, or on the way to the Falls. Through the *Clove*, one of the most remarkable mountain clefts in the world, from *Palensville* towards *Hunter*, with views of the beautiful *Fawn's Leap Fall*. To *Plauterkill* and *Stony Cloves*. To *Parker's Ledge*, overlooking the Clove. To *Moses'* and *Sunset Rocks*. To the tops of the *South Mountain*, *North Mountain*, etc. To the top of *High Peak*, laborious ascent, but with magnificent view, etc.

Division C.

CATSKILL MOUNTAINS TO AND AT ALBANY AND TROY.

Leave Catskill by rail from Oak Hill Station; or by boat from New York from Catskill Landing; making landing at

HUDSON, on the east side of the river, a large and thriving town, with considerable manufactures. [Point of departure for *Lebanon Springs* and the Shaker Village connected with them; as also for *Columbia Springs;* both minor watering-places of salubrious situation and increasing popularity. Also, railway connection east for Boston.] From Hudson, through scenery much tamer than along the Lower Hudson—past *Athens* (whence there is a railway to Albany), *Coxsackie, New Baltimore*, etc., on the left; and *Stuyvesant, Kinderhook* (residence of the late President Martin Van Buren), *Castleton,* etc., on the right—to

ALBANY, Capital of the State of New York, somewhat picturesquely situated on rising ground, on the west bank of the Hudson, with *Greenbush* opposite; the river spanned by a railway-bridge of recent erection and a certain celebrity on account of the opposition made to it by the residents of Troy, higher up the stream. It has great commercial importance, as the virtual head of sailing-vessel navigation northward; as a heavy lumber and timber depot; and especially as the point at which the immense carrying-trade of the Erie and Champlain Canals enters the Hudson.

The buildings best worth a visit and observation are the *Capitol* (soon to be replaced by a much finer erection), with the Senate and Assembly Chambers (legislative sessions from 1st January to 1st April); the *State Library*, adjoining; the *State House*, with government offices; the *Dudley Observatory*, rapidly assuming position as one of the first institutions of the kind in the country; the *State Arsenal;* the *University;* the *Medical College* (with Museum); the *City Hall; State Normal School*, &c. Rides from Albany are many and attractive—especially to the *Cemetery* (one of the handsomest in the State), to *Cohoes Falls, Lansingburgh*, and other handsome and thriving villages at practicable distance, and to some one of the *Shaker Villages* lying northward— at the latter of which (as at Lebanon), the most odd and peculiar of all forms of worship may be encountered. Leading hotels at Albany, the *Delavan, Stanwix Hall, Congress Hall*, &c.

From Albany, by street-car, omnibus or boat to

TROY, some eight miles up the rapidly-diminishing river from the Capital, where will be found nearly a rival of the latter in size and population, its superior in beauty of location, and not only a flourishing town in general manufactures, but one of the most extensive lumber and timber depots in the world. It lies on both sides of the river—the eastern portion called by the common name, and the western, *West*

Troy. There is much manufacturing, of various heavy kinds in both divisions, but especially in West Troy, where street-cars, stoves and oilcloths are among the principal articles, while at the *Watervliet Arsenal* (United States government) the founding of small arms and munitions of war is carried on very extensively. Troy has also additional prominence from the junction of the Northern, Western and Eastern lines of railway, here occurring; it has some churches of prominence (*St. John* and *St. Paul*, the principal)—the *Rensselaer Polytechnic Institute*, and the *Female Seminary*, both popular in management and extensive in influence. Two slight eminences, near the town, bear the ridiculously classical names of *Mt. Ida* and *Mt. Olympus;* and there are two pretty cemeteries—*Oakwood* and *Mount Ida*. From Troy, also, may be conveniently reached, by carriage or other conveyance, Cohoes, Lansingburgh, &c.

Division D.

NEW YORK TO ALBANY OR TROY BY NIGHT-BOAT.

Those who have before made the passage of the Hudson from New York to Albany by daylight; or those who intend to return by some day-route, and so do not wish to consume time or experience fatigue on the route northward before reaching Albany—will be able to make the transit, so far as the latter place, by night-steamers on the Hudson, leaving

New York at 6 P. M., finding luxurious accommodation for eating and sleeping, on board, and reaching Albany or Troy at so early an hour in the morning as to ensure connection with the trains for either the Northern, Western or Eastern routes.

For this transit two lines present themselves: the People's Line (New Jersey Steamboat Company), in the very large and splendid boats of which the full luxury of American river-navigation is seen; and the Hancox Line (Albany and Troy Steamboat Company), displaying less splendor though supplying strong and efficient boats, and making a specialty of reduced prices as compared with the People's Line.

Going by either of these lines, in the long days of midsummer, a considerable portion of the scenery of the lower Hudson is passed through before the disappearance of daylight; and if time at or near the full moon can be chosen, the sail under such circumstances through the Hudson Highlands affords aspects of peculiar beauty not otherwise attainable.

Division E.

ALBANY OR TROY TO AND AT TRENTON FALLS.

The New York Central Railroad will be taken at either Albany or Troy, bending westward, up the very handsome though narrow

Valley of the Mohawk, considered one of the finest in America for tracts of quiet beauty in scenery;

and often within sight of that wonderful enterprise in original construction and present capacity of conveyance, the

Erie Canal, which crosses the whole State between Lake Erie, at Buffalo, and the Hudson, at Albany;—by *Schenectady,* a quiet little old town, principally celebrated as having been the scene of a dreadful conflagration and massacre by the Indians, during the Revolutionary War. [Railway branches here for Saratoga, Lake George, Lake Champlain and Montreal, for those who prefer.] From Schenectady, by minor stations of *Fonda ; Palatine Bridge* [point of disembarkation for *Sharon Springs,* reached hence by coach]; *Fort Plain* [whence coach conveyance to *Otsego Lake, Cooperstown* (residence of the late Fenimore Cooper) and *Cherry Valley*]; *Little Falls* (where particular attention is due to the wondrous river-and-rock scenery of the pass on the left); and *Herkimer*—to

UTICA, one of the flourishing large towns of Central New York, and Capital of Oneida County. It is pleasantly situated on rising ground on the south side of the Mohawk River, and is surrounded by very fertile lands, from which proceeds, at the hands of the Welsh and other residents, one of the principal cheese-manufactures of the country. The town stands on the site of old Fort Schuyler, of Revolutionary fame; is an entrepot of both the New York Central Railroad and Erie Canal; and has a peculiar though melancholy attraction in the large and well-managed

State Lunatic Asylum. Drives around Utica are numerous and excellent. Prominent hotels at Utica, *Baggs'* and the *American*.

Lay over at Utica one day or more, and proceed, either by carriage direct, or by cars of the Utica and Black River Railroad to *South Trenton* and thence by omnibus, to

TRENTON FALLS, on West Canada Creek, branch of the Mohawk River—a series of cascades unexcelled in the world for picturesque beauty. The principal falls are five in number, successively, passing up the stream, the *Sherman Fall, High Fall, Mill-Dam Fall, Alhambra Fall* and *Rocky Heart*. To appreciate and enjoy them thoroughly, the tourist needs to descend the bank, by stairway, to the rocky level at the bottom, as far as practicable, and pass up along the left bank, on an irregular line of shelf-path, easily found, and presenting little difficulty and no danger to the careful. The rock-strata of this remarkable gorge will excite mingled wonder and admiration,—as will the really unique collection of fossils and crystals found in the neigborhood and kept on view at *Moore's Hotel*, near the Falls. Returning from the extreme point reached, to below the Mill-Dam Fall, the stairway should be ascended, to the *Rural Retreat*, to view the High Fall from above—and way taken back to the Hotel through the fine woods. Return to Utica for pursuance of route northward.

Division F.

TRENTON FALLS AND UTICA TO NIAGARA FALLS.

Leave Utica by rail on New York Central Railroad, to

ROME, a thriving town, also on the Mohawk River and the Erie Canal. [Here, those who wish to proceed more directly to the St. Lawrence and Canada, may take Rome, Watertown and Ogdensburgh Railroad, to *Watertown,* for crossing to Kingston and the Grand Trunk Line in either direction— or to *Ogdensburgh,* for crossing to Prescott and nearest route to Ottawa]. Rome, continuing by New York Central, to

SYRACUSE, large and flourishing town of Onondaga County, at the junction of the Erie and Oswego Canals, with an immense production of salt from the Salt-wells, and the peculiar celebrity of having long been the favorite place for political conventions. It is pleasantly situated at the south end of Onondaga Lake. [Railway connection, here, southward by the Syracuse and Binghamton Railroad, to *Binghamton* and the Erie Railway; and northward to *Oswego,* on the shore of Lake Ontario, with steamer connection to Canadian ports and down the St. Lawrence. Branch line of the New York Central may also be taken, at Syracuse, direct to *Buffalo,* by

AUBURN, flourishing town on Cayuga Lake, and capital of Cayuga County, where one of the New

York State Prisons is located, and where Secretary Seward has long resided—by *Cayuga, Geneva, Canandaigua* (whence branch lines to *Rochester* and southward to the Erie Railway at *Elmira*), *Caledonia, LeRoy* and *Batavia.*]

By main line, from Syracuse, by *Clyde, Lyons* and *Palmyra,* to

ROCHESTER, on the Genesee River, one of the largest towns of Northern New York, and one of the most prosperous. It has a great natural curiosity, in *Genesee Falls,* a single cataract of eminence, in jumping from which "Sam Patch," the leaper, lost his life, many years ago. Artificially, its leading attractions are the great *Erie Canal Aqueduct* over the Genesee; the *Rochester University* and *Theological Seminary; Mount Hope Cemetery; St. Mary's Hospital,* etc. [Railway connection southward to the Erie Railway, at *Corning;* also by rail to *Charlotte,* on the lake shore, whence boats to all points on Lake Ontario]. Leading hotels, the *Osborn, Congress, Brackett,* &c.

From Rochester, by New York Central, by *Brockport, Albion, Medina,* and *Lockport* (point of entrance into the Erie Canal, from Lake Erie), to *Niagara* (village), and

NIAGARA FALLS, first natural curiosity of America and admittedly among the first in the world.

Division G.

AT AND ABOUT NIAGARA.

Most students of geography, even those who have never traveled, know that the Falls of Niagara lie between the State of New York, and Canada, and that they are formed by the rushing through the comparatively narrow pass of the Niagara River, over a curved shelf of uneven rocks, of all the mighty mass of water going eastward from Lake Erie to Lake Ontario; and to a smaller number of non-visitors are known the additional facts that the *Horse-Shoe Fall* (Canadian side) is 1 800 feet across; that *Goat Island*, separating the two, is 500 feet in width; that the *American Fall* is only 900 feet in width; that the average depth of descent is estimated to be about 160 feet; and that the enormous amount of 100,000,000 tons of water is believed to pass over the ledge every hour—nearly 1,500,000 tons every minute, and about 250,000 tons every second or beat of the pulse! Beyond this, no additional statistics need be given, except that the banks of the river, below the falls, have a perpendicular height of about 180 feet, and that the mass of water, below, all the way to the *Whirlpool*, is compressed into an average space of about 480 feet of width.

It is scarcely necessary to say that days of sojourn at the Falls are desirable, to see them in all their varying aspect and become fully acquainted with their beauty (often underrated) as well their grandeur. The short-trip traveler, however, will be better served than otherwise, with a brief statement of the points of view most absolutely necessary and most conveniently attained. Of these are (1) that

Over the Rapids—view caught in passing from the neighborhood of the Cataract House, by the fragile-looking but perfectly-secure bridge, to Goat Island. It is doubtful whether the cataract itself is more impressive than this mad rush of waters, threatening to sweep away the beholder at any instant, and suggesting all the images of beautiful rage and fury. (2),

From Goat Island, over the Canadian Fall, the Canada shore and the lower rapids—with the shape of the horseshoe fully defined, and the rainbow almost constant during fine weather. (3),

From Terrapin Tower (small tower at the edge of the Canadian Fall, reached by bridge from Goat Island), giving the opportunity to look *almost perpendicularly down the cataract*, with other points of view nearly the same as from Goat Island. (4),

From Prospect Point, near the International House, on the American side, giving the American Fall almost at the feet, and the Canadian Fall and shore broadly opposite. (5),

From under the American Fall, down-river side, reaching that point by descent of steps or *Inclined Railway*, from Prospect Point. From no other point of view can the impression of the broken bright water really falling from the clouds, be caught in such enchanting perfection. (6),

From the River, crossing the lower rapids by boat, and looking up to the Falls from the greatest attainable depth below them. (7),

From the Suspension Bridges, especially the upper and smaller one, near the Falls. (8),

From the Clifton Ledge, in front of the Clifton House, on the Canadian side—the American Fall being seen from this point to perhaps even better advantage, and the whole ensemble of the Falls better caught, than even in the view (9),

From Table Rock, higher up on the Canadian side, immediately at the verge and edge of the Horse-Shoe Fall, always a favorite with experienced visitors, and from which point the view in Church's great picture was taken. Descent

Under the Falls may be made, by those who have taste for that style of adventure—either by going down the Biddle Staircase, from Goat Island (under American Fall and to the *Cave of the Winds*), or the staircase at Table Rock (under Canadian Fall, to *Termination Rock*). Neither of these descents should be made, however, without due preparation of waterproof clothing (kept on hand at both points named), and the services of a capital guide.

Lunar Island, joined by a bridge to Goat Island on the right, should be visited, in sunlight to see the *Rainbow* of the Falls in greatest perfection and, in moonlight, if the time of visit so serves, in the chance of seeing that most wonderful of spectacles, the *Lunar rainbow*.

Other Spots to be profitably visited at and near the Falls, may be named

Grand Island, very large island, above (reached by ferry) notable as the spot where Major Mordecai M. Noah, of New York, some fifty years ago commenced to build what he believed to be the City of Restoration of the Jews. (Monument commemorative, still remaining);

The Whirlpool, three miles below the Falls on American side, and showing one of the most terrible circular rushes of water in the world—with its pendant, still below, *the Devil's Hole;* the

Burning Spring, within a short walk above the Falls, on the Canada side, showing some rare phenomena in liquid combustion;

Lundy's Lane (Canada side—carriage), scene of the Battle of Chippewa (1812), with an observatory and many stories of that battle;

Queenston and Lewiston, opposite towns on the Niagara River, seven or eight miles below the Falls; the former (Canada side) with a handsome monument to the English General Brock, who fell here in 1812. Prominent hotels at Niagara, the *Cataract*, *International*, and *Park Place*, on the American side; and the *Clifton*, on the Canada side.

[From Niagara (Suspension Bridge) through Canada, by Great Western Railway, to Detroit, Chicago, and the West (including California); or, Niagara to Buffalo, and West by the Lake Shore Railroad; or, by the Grand Trunk, to Toronto,

Ottawa, Montreal, and other Canadian cities; or, rail to Kingston, and thence boat to and down the River St. Lawrence to Montreal, etc. [See Canadian routes.]

ROUTE NO. 2.—NORTHERN.

NEW YORK TO BUFFALO, NIAGARA FALLS AND CANADA, BY THE ERIE RAILWAY.

Leave New York (by morning train, for enjoyment of Delaware and Susquehanna scenery) by ferry from foot Chambers St., or foot 23d St., to *Long Dock* at Pavonia (New Jersey), midway between the towns of Hoboken and Jersey City—the immense range of wharf commanding admiration for the enterprize which has created the whole from tide-water and useless marsh; and its importance added to by its late selection as the site of the piers and houses of the White Star Line of Steamers to Liverpool.

From Long Dock, by rail, on the Erie Railway; the first point of interest after departure being the

Bergen Tunnel, through the West Bergen Hills, reached within a few moments after leaving the wharf, some three miles in length, and considered a most costly and elaborate piece of engineering, until dwarfed by recent examples in the same line. Beyond, the first town of any importance passed through, is

PATERSON, New Jersey, capital of Passaic County, in that State; the town presenting many interesting -

features in manufactures and industry. Paper, cotton, silk and other fabrics are extensively produced; and iron and steel working have even more prominence. The Ivanhoe Paper Mills, here, are the most extensive in the country; Paterson foundries boast of being able to produce steamship-shafts and other heavy irons, of greater size than any others in America; and two of the most successful and notable of the establishments for the manufacture of locomotive engines, in the world, are located here—those of Grant, and of the Rogers Co., of whom the former won the great gold medal at the French Exposition of 1867, for the splendid locomotive "America." Within the boundaries of the town are also to be seen the

Passaic Falls, on the river of the same name—well worthy the tourist's attention, from the peculiar character of the chasm into which the river leaps, and the rock-scenery in the vicinity. Beyond Paterson, the scenery, which has so far been tame, roughens and becomes better worthy of notice, as the hills of Orange County begin to break into view; and thenceforth, for a long distance, it may be said that the Erie road is one of the most picturesque in America—a marvel of wild natural beauty in surroundings, as well as of enterprize in engineering. A

Suffern's Station [junction with the old road, now used for freight only, to Piermont, on the Hudson]. commences the fine scenery of the

Ramapo Mountains, Gap and *Valley*, scene of many of General Washington's warlike operations; and the country around and beyond, entering Orange County, in the State of New York, equally celebrated for the wonderful richness of its dairy products—the noted *Orange County milk and butter*.

At *Sloatsburg*, stage may be taken to *Greenwood Lake*, a rural summer resort of much beauty and some popularity.

At *Greycourt* occurs the junction with another and now more important branch of the line—that to Newburgh, on the Hudson, by Warwick. Passengers for Greenwood Lake also proceed from Greycourt. The most important of the other stations passed on this portion of the route, is that of

Middletown, capital of Orange County, with much industry, extensive iron-works, an academy, and a surrounding country at once fertile and picturesque. Beyond Middletown soon comes into view the magnificent scenery and bold engineering operations connected with the great

Shawangunk Mountain, the passage around which, by railway, was once deemed impossible. From this point, alternate rock cuttings of great depth and length, and magnificent views over the Neversink Valley and into the wild gorges of the Upper Delaware (river), of which the first comprehensive views are caught shortly before reaching

Port Jervis, a village picturesquely situated among the mountains, at the point of junction of three

States—New York, New Jersey and Pennsylvania, and once enjoying evil repute from the facility with which doubtful characters residing there could quickly change their State and thus baffle the officers of justice. It is now a place of limited summer resort and the end of the first or Eastern Division of the Erie road. [*Falls of the Sawkill*, fine cascades, six miles distant, by carriage or stage-coach.]

Beyond Port Jervis the tourist enjoys fine views of the Delaware and Hudson Canal, in full operation; and then comes the yet wilder scenery of the Upper Delaware, the road running in many places closely along its high rocky banks, and the engineering of the whole line at this section worthy of being remembered beside that of the Rhone Valley road among the heights of Jura, and that of the road through the Apennines between Bologna and Florence. At near *Shohola*, perhaps the finest and wildest portion of the railway scenery is passed; though the views approaching and leaving *Lackawaxen* should by no means be lost. Passing Mast Hope, Narrowsburg, Callicoon (the latter and indeed all the places lately named, great headquarters for trout-fishermen and mountain-sportsmen generally) and Hancock,

At *Deposit* dinner is found, and farewell is bidden to the Delaware River. Beyond this point the grade is somewhat heavy and the ascent slow, until the top of the ridge is reached, after which follows correspondingly rapid descent for a certain distance. Not long after commencement of the descent, is crossed the once celebrated

Cascade Bridge, with a single arch over a ravine nearly two hundred feet in depth (now changed to a high embankment); and here begin to be caught wonderful views over the lovely Valley of the Susquehanna and the fine River of that name. Very soon after is crossed the

Starucca Viaduct, a splendid stone structure some 1,200 feet in length and about 120 feet in height—considered one of the noblest railway bridges on the Continent, while the scenery from and around it is wondrously lovely and attractive. Still another high crossing is made over a fine wooden trestle bridge, at *Lanesborough;* and then is reached

Susquehanna, an important station and the end of the second or Delaware division of the road, as well as noted for engine-work and other heavy manufactures. Only a few miles beyond is reached

Great Bend, another important railway station, and the point of intersection with this road, of the Delaware, Lackawanna and Western Road, from the Coal Regions of Pennsylvania. [Near Kirkwood, next station beyond, may be seen an old wooden house possessing a certain interest as the place of birth of the first Morman prophet, Joe Smith]. The next place of importance reached is

BINGHAMTON, handsomely situated at the junction of the Susquehanna and Chenango rivers, and deriving its name from an early settler, Mr. Bingham, ancestor on one side of the present noble English banking family, the Ashburton Barings. It has of late years enjoyed the distinction of being the site of the *New York State Inebriate Asylum*, of which a view of the handsome and extensive buildings can be caught from the train, on the right. Binghamton is very thriving, and considered very beautiful, healthy, and well worth a brief sojourn for examination. [Connection, here, with the Central Road, by the Syracuse and Binghamton Railroad]. Beyond Binghamton, the next important station is

Owego, a large and handsome village, on a creek of the same name, near which, at a short distance, may be seen *Glenmary*, residence once occupied by the late N. P. Willis. [Connection, here, northward, by branch railway, to Ithaca, on Cayuga Lake]. Some half dozen stations beyond, is reached

ELMIRA, another chief town of Western New York, lying on the Chemung River, surrounded by handsome scenery and displaying much thrift and prosperity. [Connection, here, northward, directly with Niagara Falls by the Elmira, Canandaigua and Niagara Falls Railway; and southward to Harrisburg, Philadelphia, etc., by the Williamsport and Elmira Pennsylvania Central, and other intersecting lines].

At *Corning*, also on the Chemung, occurs a peculiar connection with the coal-fields of Pennsylvania, by the Corning and Blossburg Railroad. Also, a branch of the Erie road runs northward direct to Rochester.

At *Hornellsville* the Erie Railway branches into two main lines, the one leading west, by

Salamanca [junction with the Atlantic and Great Western road, southwestward to Corry and the Oil Regions of Pennsylvania] and Dayton, to

DUNKIRK, on Lake Erie, terminus of the Erie Railway in that direction, and point of junction with the Lake Shore Railroad for Cleveland and Toledo (Ohio), Chicago, and other points west and northwest.

The second or northern branch of the Erie road, leaving Hornellsville, runs northwestward, by Nunda, *Portage* (with splendid bridge, of great height, and fine Fall of the Genesee, called *Portage Fall*), Warsaw and Attica, to

BUFFALO, on Lake Erie, largest town of Western New York, and one of the most important commercial depots of the Middle States. It only dates from the commencement of the century, owing much of its rapid early progress to the enterprising and unfortunate Benjamin Rathbun, who involved himself fatally in the attempt to make it the Queen of the Lakes. It is the point of entrance to the Erie Canal, from the Lake, and enjoys an immense grain and other shipping trade with the West, by steamers and large schooners. It has now not less than 40 large grain warehouses, with capacity for storing six to eight millions of bushels; has very large iron manufactures; has several public grounds —Terrace Park, Niagara, Delaware, Washington,

Franklin and other squares; has a University, Medical School, Orphan Asylum, Marine Hospital, &c.; and many of the public buildings, including the City Hall, Custom House, Post Office, State Arsenal, Market Houses and some of the Churches (the Roman Catholic Cathedral especially) are worthy the attention of even the temporary sojourner. Afternoon breezes from the Lake, facilities for water-excursions, proximity to the Canadian shore, &c., make Buffalo a charming place of abode during the hot season, though the atmosphere is often too damp for the health of invalids inclined to pulmonary trouble. Prominent Hotels, *Mansion House*, *Courter House*, *Western*, *Genesee House*, *Revere House*, &c.

[Buffalo, by rail to Niagara Falls and Suspension Bridge, for Canada and the East, or for the West, (See close of previous route). Or, direct to Sarnia, Detroit, Chicago, &c., by the Grand Trunk Railway. Or, to Dunkirk and the Lake Shore road thence to the West. Or, by Lake steamer to Cleveland and other points westward.]

ROUTE NO. 3.—NORTHERN.

NEW YORK TO SARATOGA, LAKE GEORGE, LAKE CHAMPLAIN AND MONTREAL, WITH OPTION OF THE WHITE MOUNTAINS.

New York to Albany or Troy, as by Northern Route No. 1. Thence train on the Rensselaer and Saratoga Railroad along the Hudson and Mohawk Rivers, and with a view in passing of the *Falls of Cohoes*, on the latter, and also of the Erie Canal and of Round Lake—to

Ballston Spa, once the rival of Saratoga as a place of medicinal and fashionable resort, and still frequented by a considerable number of health-seekers, though the largest of the hotels, the *Sans Souci*, has long since been converted into a seminary, and the tide of summer travel has turned towards the more celebrated springs. Ballston has original advantages of location over Saratoga, the fine creek or small river, the Kayederosseras, flowing through it, and materially adding to pleasantness as an abode; and while as a watering-place it will never quite decay, it may some day see a return to its old popularity. From Ballston, half an hour, through very flat though well shaded country, to

SARATOGA (better known as "Saratoga Springs") —the most fashionable of the American Spas.

Division A.

AT AND ABOUT SARATOGA.

This most celebrated of summer resorts on the Western Continent, with the possible exception of Niagara—has few natural features to produce such continued celebrity, its situation being comparatively low, its soil sandy, and its climate decidedly hot in midsummer. But long care and much expense have made its grounds shaded and attractive; and the number and varied character of its springs have counterbalanced all opposition and given it a popularity not likely to lessen during the present century. During the past few years, speculation (not to call it by any worse name), has joined with liberal enterprize in providing extraordinary attractions, in the shape of

The Race-Course absorbing attention during a certain number of days of the season, and presenting some of the worst features of the English turf, in the way of high betting; and

The Play-House, in imitation of Baden-Baden and Hombourg, with the addition of being owned and managed by an Honorable M. C. Another and more meritorious feature is

The Leland Opera House, near and attached to the Union Hotel, and affording splendid opportunities for concerts, grand balls and other festivals, more pleasant to the sojourners than (it is to be feared) profitable to those providing the accommodation. Of course the principal source of popularity and profit has been found in

The Springs, of which the whole number must approach twenty, very different in character, while upon two or three of them has been concentrated, until lately, nearly the whole popularity giving patronage to the group. The waters of the *Congress* lead the list, now, instead of monopolizing as they once did: they are bottled extensively and sent everywhere, as well as consumed unlimitedly at the Spring. After them, of late, have come the *Empire*, pressing close upon the Congress as an article of commerce; and no small amount of popularity in the same line is being attained by those of the *High Rock* (held to be specially strong and medicinal), the *Hathorn*, the *Constitution*, &c.,—while the *Iodine*, the *Columbian* and others command extensive home-consumption. The virtual "Pump-Room" of Saratoga, meanwhile, has been and continues in the Congress Spring, most picturesequely located and best kept, and with fine grounds near to add to its attraction.

[Saratoga suffered very severely by fire in 1865 and 1866, two of the oldest and largest of the hotels, the United States and Congress Hall, almost as truly features of the place as the Springs themselves, and endeared by a thousand recollections as well as made classic by Willis' charming sketches, going down in those years. One of them has since been rebuilt, however, with enlarged accommodation; the other will be; and although a large supplementary fire, in September, 1871, destroyed several minor houses, there is no fear whatever of Saratoga permanently suffering from deficiency of hotel accommodation.]

[The hint is worth something, to strangers—that the most delightful time for visiting Saratoga is to be found later than the full season—say in September and early October, when the climate is delicious, and when the loveliest sunsets of the world (finer than the Italian) can be enjoyed from the hills just northward.]

Excursions from Saratoga are not many or remarkably varied. The most popular is found in the afternoon ride to the *Lake*, six miles distant—a handsome forest-girded bit of water, with fine facilities for boating and fishing, and with Moon's and Abell's "Lake Houses" to supply entertainment to visitors. With this ride is often combined a visit to *Barhydt's Mill*, on the road—also made classic by Willis. Another excursion, generally made, is that to the *Battle Field of Stillwater*—scene of Sir John Burgoyne's surrender to Genl. Gates in 1777.

Prominent Hotels at Saratoga, *Union Hall*, *Congress Hall*, *Clarendon*, &c.

Division B.

SARATOGA TO AND AT LAKE GEORGE.

The route from Saratoga is by cars of the Saratoga and Washington Railroad, to

Moreau Station, where stage-coach is taken for the crossing of the intervening mountains. This ride is a notable one, the views of the Upper Hudson, Fort Edward, and the different ranges of hills within view, keeping the traveler continually on the alert, especially if the weather allows an outside seat on the coach. Within a few miles from Moreau is passed through

Glenn's Falls, on the Hudson, the village most wildly and picturesquely located at a pass of the river through rocks of terrible roughness, and the interest materially added to by the fact that among these broken rocks Cooper laid the scene of a part of the "Last of the Mohicans." Half way between Glenn's and the Lake is passed

Bloody Pond, scene of the defeat of Col. Williams by the French and Indians under Baron Dieskau, and slaughter of all his force, in 1755, during the old French wars, and of which the name is said to have been derived from the literal filling of the pond with bodies on that occasion. Shortly after leaving Bloody Pond, and on emergence from the forest on high ground, is enjoyed one of the most magnificent views supplied by the Western Continent, in the beautiful "Horicon" (Indian name of Lake George) its islands and mountain borderings.

Caldwell, south end of the Lake, is the spot where the traveler is set down, from Moreau; and here and near are located some of the most enjoyable of the hotels welcoming visitors

Rowing, Sailing, and Fishing on Lake George are privileges not to be ignored: the more enjoyable for the wonderful clearness of the water, which often allows the bottom to be seen at twenty or thirty feet, and which won for it from the French the name of "Le Lac du St. Sacrament" and induced the carrying of the water to great distances for baptismal purposes. The variety of fish caught—trout, perch, pike, &c., commends it to the special favor of sportsmen. It is while on the water, too, that the beautiful panorama of the Lake, with its islands and encircling mountains, admitted to be among the finest on the globe, can best be enjoyed.

The Old Forts must command a certain degree of attention from the visitor to Lake George, combining, as they do, historical interest with their location. Of Fort William Henry, the small traces of embankment remaining lie immediately beside the Hotel of the same name, from in front of which the best views are commanded and the steamboat embarkations on the Lake are made. Of Fort George, half a mile eastward, considerable portions of the crumbling walls yet present themselves, half buried by earth and overgrown with trees. Leading hotels at Lake George, the *Lake House* and *Fort William Henry.*

[From Lake George may be visited, by stage-wagon or other conveyance, *Schroon Lake* a small wild, picturesque sheet of water lying north-westward; and thence, the lower part of the

Adirondack Mountains among the boldest and most interesting chains of the East, and of late years very popular as resorts for pleasure seekers and health-seekers who have no objection to "camping-out" and "roughing" it a little. Particulars of special routes, best obtained of those who manage the conveyances; as all other information on such partially-opened lines, must be more or less unreliable and risky.]

Leaving Caldwell for the passage northward, a small steamer is taken; and the entire passage to Ticonderoga, about 35 miles, is one of the most charming in any land, affording otherwise unattainable views of the surrounding mountains and the almost countless islands of the little body of water so favored, the actual number of which is said to reach nearly or quite three hundred. Among the most notable of these in this part of the Lake, are *Diamond Island*, Burgoyne's military depot in 1777; *Long Island*, north of Diamond; *Twelve Mile Island*, near Bolton. Not far beyond the latter, *Tongue Mountain* thrusts itself out into the Lake to a great distance (whence the name), forming a part of the *Narrows*, entered just beyond, under the shadow of *Black Mountain*, the highest peak of the lake-shore. North of the Narrows comes *Sabbath Day Point*, a strip of low, cultivated land, so named, as alleged, by Genl. Abercrombie, from a Sunday morning embarkation of troops made there. Not far beyond, passing the bold headland of *Rogers'*

Slide, are passed *Prisoners' Island,* used as a place of military confinement by the English during the old French war; and *Lord Howe's Point,* where that English general landed to make his attack on Ticonderoga. Directly beyond comes an insignificant landing, at which concludes the beautiful sail on Lake George, and the tourist enjoys a three mile ride, often in a rough wagon instead of a stage-coach, over a rough road that still seems to be unobjectionable and in keeping with the journey, to the ruins of Ticonderoga.

Division C.

TICONDEROGA BY LAKE CHAMPLAIN TO MONTREAL.

Crossing from Lake George, by stage-coach or wagon, as above,

Fort Ticonderoga will be found only a ruin, with one gable remaining to show its original height. It was first constructed by the French, about 1756, but derives its principal interest from the peculiar mode of its summons to surrender by the madcap Vermonter, Ethan Allen, in 1775, the formula being: "In the name of the Continental Congress and the Great Jehovah!" It was soon retaken by the British, however, and held during the war. [From Ticonderoga detour may be made to *Crown Point,* another place of revolutionary interest, taken at nearly the same time; and thence may be reached most conveniently by wagon and on foot, *Lake Sanford, Lake Hender-*

son, and beyond the famous *Indian Pass* and the great peaks of the *Adirondack Mountains; Tahawus*, or *Mount Marcy, Mount McIntyre*, the *Dial Mountain*, etc.]

At the wharf at Ticonderoga, to continue main route northward, steamer on Lake Champlain is taken; and thenceforth, in fine weather, is found a sail of many hours, not often equaled in enjoyment. Besides the towns and hamlets studding the shores, there are special points of interest on the Lake in the shape of singular rocks and islands, of which the most notable, below, may be mentioned as *Split Rock*, an immense mass of ironstone, half an acre in extent, split away from the main only about twelve feet; the *Four Brothers*, small islands always haunted and half covered with noisy gulls, like Ailsa Craig; *Juniper Island* and *Rock Dundee*, both masses of rocks rising to the height of over 30 feet. It is not in these particulars, however, that lies the chief charm of sailing on noble Champlain (130 miles in length, and width varying from $\frac{1}{4}$ mile to 13: area covered, about 500 sq. miles). That principal charm lies in the presence of a noble range of mountains at either side: at the right, the *Green Mountains* of Vermont, among the highest in Eastern America; on the left, at greater distance, the equally noble range of the *Adirondacks*.

Landings are made, on the left, at

Plattsburg, New York, a thriving town and scene of the land-and-naval battle between the English and Americans in September, 1814. [From **Plattsburg** may be visited *Keeseville*, a thriving town of Essex County; and thence the *Au Sable River*, the *Saranac Lakes*, and the great fishing and pleasure grounds of the NORTHERN ADIRONDACKS, may all be reached by taking stage from Keeseville and tracing out the special routes through experience and information locally derived. Certain provisions and conveniences, however, will be necessary before leaving civilization for the wilderness; and these should be provided, after obtaining the special local advice, before leaving the towns on the shores of Lake Champlain.]

Important stoppage, on the right, is made at

BURLINGTON, Vermont, one of the largest and most important towns in the State, with a University, many fine buildings, manufactures, and a great concentration of railway, steamboat and stage-coach routes. At Burlington are also enjoyed, as from the Lake approaching it, fine views of the two highest peaks of the Green Mountains, *Mt. Mansfield* and *Camel's Hump*. Hotels, the *American*, &c.

Landing is made from the steamer, 120 miles from Whitehall, at

Rouse's Point, an unimportant place except for this transit, lying within the United States but on the very border of Canada. Thence railway, (branch of the Grand Trunk) by *St. John* to *La Prairie*, on the St. Lawrence, and ferry to Montreal.

[The tourist, who, after visiting Saratoga lacks

ROUTE NO. 3.—NORTHERN.

time for Lake George and Champlain on his way to Montreal and other Canadian cities,—leaving Saratoga, by the Saratoga and Washington railroad, instead of stopping at Moreau Station, will continue by rail to

Whitehall, at the southern end of Lake Champlain—a town of much importance in connection with the lumber trade and the extensive transportation through the Champlain Canal, commencing there, from the Lake to Albany. Thence continuing by rail, by *Castleton ;* by

Rutland, thriving town of Vermont, with railway connections in all directions, pleasant location and fine mountain-and-valley scenery (well worthy of a short sojourn); by *Middlebury*, by *Burlington* (noticed in Champlain route); by

St. Alban's, one of the handsomest towns in Vermont, lying near Champlain, with fine scenery, salubrious air, many rich buildings, and an immense butter-and-cheese market; thence by Rouse's Point and St. John's, to La Prairie and Montreal.]

(Excellent hotel at St. Alban's, the *Weldon House.*)

[The tourist who wishes to visit the White Mountains from this direction, with or without proceeding to Canada can do so by either of the three following routes, after having visited Saratoga and Lake George:

(1) Proceeding to Burlington by boat on Lake Champlain, as before arranged, land there and take rail by Essex Junction and *Wells River*, to *Mere-*

dith Village, New Hampshire, on the shore of Lake Winnepisaukee, thence to *Centre Harbor*, on the Lake, whence rail and coach to *Conway*, for the White Range. Or, (2) land at Burlington, proceed by rail to *Wells River* and thence to *Littleton*, for the Franconia Range. Or, (3) if proceeding from Saratoga direct by rail, go by Whitehall, Castleton and Rutland to *White River Junction*, whence option of either of the ranges, by proceeding to Meredith or Littleton, as before just-named.]

[From Littleton to Montreal or Quebec, by the White Mountain road and the Grand Trunk, if desired, after visiting only the Franconia Range; or from Conway, after seeing the White Range, by *Yarmouth Junction* (PORTLAND, at option), or crossing to *Gorham* by carriage and thence on by rail of the Grand Trunk, to either Montreal or Quebec.]

ROUTE NO. 4.—EASTERN.

NEW YORK TO BOSTON, BY NEW HAVEN, HARTFORD, AND SPRINGFIELD (RAIL); BY PROVIDENCE (RAIL); BY NEWPORT, NEW LONDON OR STONINGTON (SOUND-BOAT).

Division A.

BY RAIL, BY SPRINGFIELD.

Leave New York by morning express of New Haven Railroad, by *William's Bridge* (point of divergence of the New York and Harlem road), by several unimportant stations (with occasional and pleasant views of Long Island Sound, at the right), to *Stamford*, *Darien* and *Norwalk* (Connecticut), (the latter the scene of a serious accident, train running into the River, at the Bridge, many years since); then by other unimportant stations, to

Bridgeport, on Long Island Sound, large and important manufacturing town, especially in the detail of Sewing Machines, of which two of the largest factories in the world, the Wheeler & Wilson, and Howe, are located here. Also, point of intersection of lines of railway leading to the manufacturing interior of Connecticut (Housatonic and Naugatuck); and old residence of the celebrated showman, P. T. **Barnum.** Bridgeport to

NEW HAVEN, also on Long Island Sound (lines of steamers to and from New York); one of the most important towns of the East, and seat of *Yale College*, as well as noted for the shaded beauty of its streets and the peculiar magnificence of its elm-tree avenues, the latter feature giving it the name of the "Elm City." New Haven is well worthy of a sojourn for examination. The first point of interest is of course to be found in the *College Buildings*, of which there are some fifteen, with a remarkable air of antiquity, for America (the College founded in 1700); the *College Green*, or *Campus*, with *Chapel*, and fine old *Churches;* the *State House*, an imposing edifice, of marble, recently completed; the *Fine Art Building* and *Trumbull Gallery*, in the latter of which are the original fine historical pictures by Colonel Trumbull (Washington, etc.); the recently built *City Hall*, etc. Some interesting Monuments are to be found in the *Grove Street Cemetery;* among others those of Roger Sherman, founder of Rhode Island, Noah Webster, Pierrepont Edwards, the poet Hillhouse, etc.; and in the Campus a lonely tomb, with inscription, is that of the English regicide Goffe, who fled to America, with Whalley, and died here. Some interesting excursions are to be made from New Haven: to

East Rock, a remarkable eminence at two or three miles distance, and a favorite resort—to *Savin Rock*, bathing place on the Sound; and, more distant, to *Wallingford* and *Hanging Hills*.

Prominent hotels at New Haven, the *New Haven House*, *Tontine*, *Tremont*, &c.

Resuming rail, on the New Haven, Hartford and Springfield road, and now running northward, away from the Sound—passing *Wallingford*, *Meriden* (great Brittaniaware and other metallic manufactory) and other stations of less importance, is reached

HARTFORD, on the Connecticut River, a large town with much beauty of location, large manufacturing interests, and dividing with New Haven the seat of government of the State. (Reached by steamers from New York). It has an educational institution of eminence, known as *Trinity College;* the *Connecticut Historical Society;* the *Watkinson Libary*, with rare books, pictures and statuary; *Wadsworth Atheneum,* etc.; and among the handsome buildings of the town are the *Deaf and Dumb Asylum, Retreat for the Insane, Hospital,* etc. The *Charter Oak* (place of hiding of the old Connecticut Charter from a tyrannical Governor) long one of the boasts of Hartford, blew down in 1856; but the place where it stood is still shown by a slab. Among present curiosities are the late *Col. Colt's Fire-Arms Manufactory*, the house occupied by the late Mrs. Sigourney, the poetess, etc. There are also many picturesque short excursions from Hartford, the most prominent among them being those to *Talcott Mountain, Wethersfield, Prospect Hill,* &c. Leading hotels, *Allyn House* and *United States*.

Beyond Hartford, pasing *Windsor*, and *Windsor*

Locks (water power and important manufactures), and other minor stations, is reached

SPRINGFIELD, Massachusetts, also lying on the Connecticut River, one of the most important towns of the State, and peculiarly notable for its diversity of railway communication. [Connection, here, eastward to Boston and westward to Albany, by the Boston and Albany road; southward to Hartford, New Haven and New York, by the New Haven, Hartford and Springfield; northward to the White Mountains of New Hampshire, to Vermont and Canada, by the Connecticut River and other intersecting roads]. A prominent source of prosperity as well as object of interest is the *United States Arsenal and Arms Manufactory*, the largest in America, located here, furnishing extensive employment and supplying the celebrated "Springfield Rifle." (Written of by Mr. Longfellow, in one of his finest poems, "The Arsenal at Springfield.") It has also other and important manufactures—especially of carriages; handsome public grounds, in the *Cemetery, Hampden Park*, &c.; and is cons'dered one of the most charming places of residence in the East. Prominent hotels, the *Massasoit, Cooley's*, &c.

From Springfield, by *Palmer*, direct to

WORCESTER, another of the large and important towns of Massachusetts, with extensive manufactures and even more numerous railway connections than Springfield. [Westward, by Boston and Albany road (Western) to Springfield and Albany, and Springfield, New Haven and New York. Eastward, by Boston and Worcester, to Boston. Northward, by Worcester and Nashua, to the White Mountains and Canada. Southward, by Norwich and Worcester, to New London, and Shore Line road and line of steamers to New York. South-eastward, by Worcester and Providence, to Providence, Newport, &c.] It has some fine public buildings, among which may be mentioned the *State Lunatic Asylum, Mechanics' Hall,* building of the *American Antiquarian Society,* &c., and divides with Springfield pre-eminence in the manufacture of railway and other carriages. Leading hotel, the *Bay State.*

From Worcester, by *Grafton, Framingham, Natick* (home of U. S. Senator Wilson) *West Newton, Brighton,* and other minor stations, direct to

BOSTON. [See ahead—" At and about Boston."]

Division B.

BY RAIL, BY PROVIDENCE (SHORE LINE).

Leave New York by 12.15 P. M. express of the New Haven Railroad. (May leave by night-express, but at sacrifice of scenery). Take tickets by "Shore Line," as distinguished from those by "Springfield;" and be sure that Shore Line carriage is taken To *New Haven,* as from Springfield route preceding.

From New Haven, due east (New Haven, New London, and Stonington road) along or near the shore of the Sound, with closer and still finer views than those before reaching New Haven—by Guilford, Madison, Clinton &c., to

Crossing of the Connecticut River, at *Lyme*. (Bridge, formerly ferry-boat, carrying part of the train, with through passengers). Then by S. Lyme and E. Lyme, Waterford, &c., to

NEW LONDON, at the mouth of the Thames River —a sea-port of some importance, and formerly depot of one of the great whaling-fleets, before the late decay of that trade. [Line of large and fine steamers from and to New York, every evening: communicating by rail to and from Boston, by Norwich and Worcester.] Harbor considered one of the finest on the Atlantic coast, and defended by splendid fortifications, *Fort Trumbull* being the principal, below the city and at the right, towards mouth of the harbor. New London, an attractive place of resort and residence in many respects, has a peculiar and melancholy interest as having been long the business place of Benedict Arnold, the traitor, whose sign over one of the old shops is still pointed out. The principal public buildings, worth notice, are the *Custom House, Court House, Female Academy*, &c. At the mouth of the harbor the *Pequot House* is a very attractive place of summer sea-side resort.

From New London the Thames is crossed by ferry boat (part of train carried over, with through-passengers, and lunch on boat) to *Groton*, on the opposite bank—for some years the depot of another steamboat line between New York and Boston, now

ROUTE NO. 4.—EASTERN.

abandoned. Groton, by rail, still within frequent sight of the Sound, to *Mystic*, somewhat famous for wooden-ship building, and

Stonington, Connecticut, Sound port of some importance, lying at the mouth of river of same name. [Line of large steamers to and from New York, every evening: communicating by rail to and from Boston, by Providence.] Stonington, by Westerly and other stations, including

Wickford [railroad and steamboat communication direct to NEWPORT, in connection with trains]; and

East Greenwich [steamboat direct to NEWPORT, in connection with trains]—to

PROVIDENCE, capital of the State of Rhode Island, and one of the principal towns of the Eastern States. It lies on Providence River, extending from Narragansett Bay; has much beauty in location and enormous wealth in buildings and the appointments of residents; manufactures very extensively, in engines, heavy machineries, cottons, prints, jewelries, &c.; and has nearly two hundred and fifty years of antiquity since its foundation by Roger Williams, when driven from the colony of Massachusetts Bay on account of his religious opinions. It is the seat of *Brown University*, a literary institution of the first eminence, with a large and valuable library; and is also noted for the number and excellence of its public schools. The two handsomest public buildings in the town are the *Rhode Island Hospital* and the *Custom House;* though there are very many fine edifices connected with the numerous public charities of the city, among which may be named the *Reform School*, the *Dexter Asylum for the Poor, Butler Hospital for the Insane, Home for Aged Women*, &c. It has also several imposing churches, banks and other edifices; and *Swan Point Cemetery, Narragansett Park*, and other public grounds deserve visit and notice. The *Soldiers' and Sailors' Monument*, recently erected, is also worthy of notice. Leading hotels, the *City Hotel*, and *Aldrich House*.

[Railway communication with Boston, by Boston and Providence road; with Worcester, by Providence and Worcester; with Hartford, by Hartford, Providence & Fishkill; with Newport, by Providence and Newport; with New London, by Stonington and Providence, &c.; with New York, by road just traversed, and by lines of steamers from Fall River and Newport, &c.],

[*Pawtucket*, near Providence, is the seat of heavy manufactures, and of the first cotton-mill ever built in America.]

Providence, by Pawtucket, Attleboro, Mansfield, Foxboro, Readville, &c., to

BOSTON.

Division C.

BY BOAT, BY NEWPORT OR FALL RIVER.

Leave New York, 5 P. M., daily (except late autumn, winter, and early spring, when the hour is 4 P. M.—see bills at hotels) by boats of the Narragansett Steamship Company, by Long Island Sound, for Newport or Fall River as may be preferred. This route, as well as the other Sound routes about to be named, affords not only a delightful sail, in the customary fine weather of summer, but conveys a better idea than can otherwise be attained, of the size and magnificence of the vessels employed in this transit. Leaving the pier and proceeding past the whole line of the city, then past the islands lying in the Sound or East River, and up the Sound itself,—unequalled opportunities are enjoyed for observing the waterfront of the city, the extent of the penal and benevolent institutions on *Blackwell's Island, Randall's Island, Ward's Island,* &c., the rocky dangers of *Hell Gate,* the attractive scenery of the river shores (Long Island on the right; New York or Manhattan Island on the left); the extensive fortifications guarding that approach to the city, in *Fort Schuyler,* at Throg's Neck, etc.—daylight lasting, in the warm season, until all these points of interest are passed, and the remainder of the course up the Sound offering few attractions other than those of a marine character.

At an early hour in the morning is reached the end of the route by Sound, and place of debarkation for those who wish to vist the watering place and afterwards proceed thence to Boston by rail,—in

NEWPORT, one of the largest and most important towns of Rhode Island, and one of the most celebrated and fashionable of American sea-side resorts, for the past quarter of a century or longer. It lies on Narragansett Bay, at near the entrance from Long Island Sound, and boasts a harbor of peculiar beauty as well as one of a depth of water almost unequalled. It has a fine bathing-beach, markedly safe, within short-riding-distance of the principal hotels; and at greater distance are to be found and visited the remarkable groups of rocks known as *Paradise*, *Purgatory*, the *Hanging Rocks*, etc. The *Glen*, the *Spouting Horn*, *Lily Pond* and the *Dumpling Rocks*, are also places of much picturesque interest to visitors and residents; while at some ten miles distant, south-eastward, lie *Seconnet Point* and *West Island*, the latter supplying the very best sea-shore-fishing on the American coast. *Fort Adams*, at near the mouth of the harbor, is one of the largest and strongest fortifications in the North; and the ride to it, from the town, is one of the afternoon features of Newport life. Another peculiarity of Newport is the fine sailing in and about the harbor, securing the constant presence of yachts, and many regattas during the season, on a more or less extensive scale. And yet another is to be found in the wide extent of lawned and terraced bluff, overlooking the sea, where are located an immense number of summer-cottages of the wealthy, giving a higher tone to the prevalent hotel-life, even while moderating and to some degree lessening it. One marked object of interest is to be found at Newport—the *Round Tower*, alternately called a Norse remain and a wind-mill, but around which Longfellow, adopting the former belief, wove his marvellously beautiful poem, the "Skeleton in Armor." There are also some patriotic erections and antiquities of interest: among the structures the old *State House, Commodore Perry's House,* the *Vernon Mansion,* the *First Baptist Church* (1638), the *Perry Monument,* &c.; and of minor antiquities, *Franklin's Printing Press* (now or late in the office of the *Newport Mercury* newspaper), the ancient *Chair of State* of the Colony, etc. Of modern buildings of merit, the number is considerable. Newport has the additional celebrity of having been the birth-place of Gilbert Stuart, the painter, Malbone, the miniature-painter, and Commodore Perry; and Cooper flung round it a romantic interest as the opening scene of the "Red Rover." Leading Hotels: the *Ocean House, Atlantic, United States,* and *Perry.*

[Above Newport, within convenient riding or sailing distance, lies the fine eminence of *Mount Hope,* with interesting reminiscences of the Indian King Philip, and splendid views over Narragansett Bay, the city, harbor, &c.]

[Newport to BOSTON, by rail, by Fall River and connection with the Old Colony road, or by Providence.]

If not wishing to stop at Newport, and still proceeding to Boston, on some New York steamer from which debarkation has before been supposed, the route will be pursued by remaining on boat until its next and final landing, at

FALL RIVER, a thriving manufacturing town on Taunton River, eastern branch of Narragansett Bay, within the State of Massachusetts though near the Rhode Island border. Peculiarly noted for works in machinery, in heavy irons, and in cottons and prints. [Communication with Newport and Providence by boat and rail.] *Mount Hope*, before alluded to, is in view from Fall River and may be most conveniently reached from this point; and the bridge connecting Rhode Island (island) and the main land is near, at *Tiverton*.

Fall River to BOSTON, by rail of the Old Colony Road.

Division D.

BY BOAT, BY NEW LONDON, NORWICH & WORCESTER.

Leave New York, 5 P. M., by boats of the Norwich and Worcester Line, on the Sound, the route displaying precisely the same features as that before named, (except that the run by sea is shorter and less exposed in rough weather, and that rail is taken at an earlier hour) to

NEW LONDON [see previous description.]

New London, by rail on the Norwich and Worcester road, to

NORWICH, very old and handsome small town of Connecticut, picturesquely situated at the head of navigation of the river Thames, with its steep streets literally lying on terraces, but many manufactures, much commercial prosperity, and no small number of old buildings recalling the early historic days of the State.

From Norwich, still by rail of the Norwich and Worcester road—by *Plainfield* [junction, for Providence, Newport, &c.], by Putnam, Webster, WORCESTER, &c., to BOSTON.

Division E.

BY BOAT, BY STONINGTON AND PROVIDENCE.

Leave New York, 5 P. M., by boats of the Stonington Steamboat Company, on the Sound, with same features as those of two previous routes, though less extensive in sea-voyage than that to Newport, and longer than that to New London—to

STONINGTON, Connecticut (before referred to in Shore-Line route by rail), at mouth of Stonington river, near the Sound,

Stonington by *Providence*, by rail, with same features shown in that division of the Shore-Line rail route, to BOSTON.

Division F.

AT AND ABOUT BOSTON, WITH EXCURSIONS.

BOSTON, Capital of the State of Massachusetts, one of the largest, most influential and handsomest of the cities of America, and in many regards the most remarkable of all—lies at the extreme western point of Massachusetts Bay, where that body of water is entered by the Charles River; and most of the old city is erected on a peninsula of several hundred acres, extending up from *Roxbury*, at the south, and curved around by the wide mouth of the Charles River, which thus divides from it *Cambridge* on the west, *Charlestown* on the north, and *Chelsea* and *East Boston* on the east. All these form parts of the present city, however, by means of different bridges spanning the river mouth and edge of the harbor; and the result is that Boston seems from some points of view to be almost as completely a "City of the Sea" as Venice. Northeast of it, at some miles distance, the bold headland of *Nahant* runs southward from the mainland at *Lynn*, behind Chelsea point, adding to the picturesqueness of the whole harbor, as well as aiding the several islands (*Castle Island*, fortified by Fort Independence; *Governor's Island*, Fort Winthrop; *George's Island*, Fort Warren; *Deer Island*, occupied by House of Industry and other public buildings; and others, minor in size and importance) in sheltering it from the rough winds of the east. An additional feature of the old city, or "Boston Proper," is found in three eminences or slight hills on and among which it was originally built, giving it the name of the "Tri-Montane City," and originating the name of "Tremont" so intimately connected with it. On the highest of these stands the *State House*, the whole city appearing to slope up to it, and the view, on approach, being thus rendered peculiarly impressive.

With reference to transit through and about Boston, it may be said that many of the streets, especially in the older portions of the town, are crooked and involved to a proverb, but that the street-car system is very extensive, complete and convenient, and that carriage-hire, though high in comparison to European, is less exorbitant than in New York.

Of Streets, the best worth noting are *Beacon Street*, at the top of the Common, the most fashionable; *Tremont Street*, at the bottom of the Common, blending of fashion and business; *Washington St.*, nearly parallel with the latter, southward, business centre; *State, Congress*, and other streets in the neighborhood of the Old State House, financial, law, etc. Other and newer streets and avenues, lying on the Back Bay, west of the Public Garden, are now, however, rapidly becoming fashionable and notable in that particular.

Of Public Grounds, Boston has two, within the city proper, of peculiar prominence: the *Common*, a triangular park of nearly fifty acres, sloping down from the State House, handsomely shaded, with a Pond or Lake, and in an enclosure near the middle, the celebrated *Old Elm* called the "Liberty Tree;" and the *Public Garden*, adjoining the Common on the West, with a handsome Lake, bridges, floral walks, and a fine statue of Washington, by Ball.

Of Antiquities, principally connected with the

War of the Revolution, Boston has many of interest: *Faneuil Hall,* Faneuil Hall Square, otherwise known as the "Cradle of Liberty," where early meetings of patriots were held (still used for meeting purposes, and containing some national portraits); the *Old State House,* State Street; *Brattle Street Church,* Brattle Street (with one of the round shot of the Charlestown bombardment still embedded in the front wall;) *Old South Church,* corner of Washington and Milk Streets; *Ordway Hall,* Province House Court, once the residence of the Colonial Governors; *Liberty Tree,* Boston Common; and many others of minor consequence.

Of Public Buildings, deserving attention, are the *State House* (Capitol), Beacon Street, with interesting military and other memorials, legislative chambers, etc., within, and splendid and extensive view from the roof; *Custom House,* foot of State Street; *Exchange,* State Street (Post Office below); *Court House,* Court Square; *City Hall,* School Street (colossal statue of Franklin, in front); *Massachusetts General Hospital,* Allen Street; *City Hospital,* Harrison Avenue; *Quincy Market,* adjoining Faneuil Hall; *Boston Public Library,* Boylston Street; *Masonic Temple,* corner Tremont and Boylston Streets, etc. Other erections of interest, the *Boston Water Works* (Reservoir), Derne Street; the *Bridges,* connecting the various suburbs with the city proper; the *Wharves* (Long, India, Central, Commercial, etc.), several of them of great extent and the system the best on the Continent.

Of Monuments, of course, the first place is taken by the *Bunker Hill Monument*, on Breed's Hill, site of the Revolutionary battle of the first name. It is a plain obelisk of granite, of great height, ascended from within, and from the top commanding a most extensive and magnificent view. Near it stands the *Warren Statue*, in honor of Dr. Joseph Warren, who fell in the battle. In State House, statue of *Washington*, by Chantrey; and in front of same building, bronze statues of *Daniel Webster* and *Horace Mann* (great Massachusetts organizer of education). In front of City Hall, colossal *Franklin*, before noticed. In Public Garden, equestrian *Washington*, by Thos. Ball, also before noticed.

Churches of prominence: *St. Paul's* (Episcopal), Tremont Street; *Christ Church* (date 1722), Salem Street; *Old South* (date 1730), Washington and Milk Streets; *Brattle Street* (Unitarian—date 1773), Brattle Street; *King's Chapel* (Unitarian—date 1750), Tremont and School Streets; *Park Street* (Congregational, with finest spire in the city), Park Street, facing Tremont; *Central* (Congregational: considered the handsomest in the City), Berkeley and Newbury Streets; *Trinity* (Episcopal—date 1735), Summer and Hawley Streets; *Tremont Temple* (general devotional) Tremont Street; *Immaculate Conception* (Roman Catholic), Harrison Avenue, &c., &c.

In Libraries and Literary Institutions Boston is peculiarly rich, the intellectual and educational status of the City being especially enviable. Among the more notable libraries may be mentioned the *Boston Public Library;* the *Athenæum,* (with gallery of paintings and sculpture); the *Mercantile; American Academy of Arts and Sciences; Natural History Society; Massachusetts Historical Society; State, Law, General Theological,* and others.

Commercial Buildings of much merit in architecture abound in Boston—notably on the streets near to the harbor, at the east side—on *Franklin, Commercial, Devonshire, Winthrop,* and other trade streets, and in the neighborhood of Long, Central and India wharves. They are principally of hammered granite, very solid and impressive, and convey a reminder of Liverpool and other commercial cities of the Old World, different from that of any other American city.

Principal Theatres, etc. the *Boston Theatre* (opera-house, at intervals), Washington Street; *Globe Theatre* (late Selwyn's), Washington Street; *Boston Museum,* Tremont Street; *Howard Athenæum,* Howard Street; *Music Hall,* Winter Street, near Tremont (with organ of immense size and power, second in the world), etc. Prominent Hotel Buildings (also Hotels): the *American,* Hanover Street; *Parker,* School Street; *St. James,* Newton Street; *United States,* Beach Street; *Tremont,* Tremont Street; *Revere.* Bowdoin Square, etc.

Excursions from the City, on foot, by horse-car, or carriage, include those to HARVARD UNIVERSITY, Cambridge, with 15 buildings, an Anatomical Museum, an Observatory, and the first educational rank in the Western World; to the *Washington Head Quarters* (now residence of Professor Longfellow, the poet,) also at Cambridge, with Tree, under which Washington took command of the American forces; to *Mount Auburn Cemetery*, four miles from the city, with Tower, commanding excellent view; Chapel, containing stained windows, busts of Adams, Winthrop, Story, &c.; and Spurzheim, Bowditch, and other handsome and attractive monuments in the grounds; to *Forest Hill* and *Mount Hope Cemeteries*, Roxbury; to *Woodlawn Cemetery*, near Chelsea; to *Cochituate Lake*, whence the water-supply for Boston is drawn; to *Wenham Lake*, whence is derived most of the American ice-supply for Europe; to *Fresh Pond*, a place of summer resort near Mount Auburn, etc. Those by boat or carriage will include *Nahant*, bold headland on the east of the harbor, once a fashionable watering-place, and always cool, attractive and pleasant; *Lynn*, near Nahant, famous as the head of the shoe-manufacture of the world; *Chelsea Beach*, *Swampscott* and *Phillips' Beach*, northward of Nahant; *Nantasket Beach*, south side of the harbor, etc.

From Boston, also, may be conveniently reached [by Boston and Lowell railroad],

LOWELL, large and thriving town on the Merrimac River, at its junction with the Concord. It is the largest of the American manufacturing towns, and considered the Manchester of the Western

World, not less than 50 to 60 large mills being employed in the manufacture of cottons, prints, woollens, etc., and the operative labor reaching to the number of from 14,000 to 15,000, a large majority females, of rare intelligence for their class. The *Pawtucket Falls*, near the city, furnish the water-power for all the *Mills*, many of which, with the Falls themselves and some of the principal buildings and public grounds of the town, are worth examination. Leading Hotels: the *Merrimac, Washington*, and *American*. [Rail connection to Groton for all points westward; to *Nashua*, for points northward; to *Lawrence*, eastward, etc.]

May also be conveniently reached from Boston, south-eastward, [Old Colony road],

PLYMOUTH ("Plymouth Rock"), place of landing of the Pilgrim Fathers, with *Pilgrims' Hall* and many interesting relics of the early settlement; and

NEW BEDFORD, on Vineyard Sound, at the head of what remains of the whaling business, and a seaport of picturesque location and prominence; or [by same and Cape Cod railroad],

Cape Cod, Yarmouth, Hyannis, and all that wild and desolate but interesting section of the Atlantic coast.

May also be conveniently reached from Boston, (by Eastern railway, by Somerville, South Malden, Chelsea, Lynn and Swampscott),

SALEM, very old town and port on the coast, with a certain celebrity on account of the witch-burnings and other events of Colonial times, and very picturesque in location and many of its buildings; but much more notable, now, as at one time the residence of Nathaniel Hawthorne and scene of his "House of the Seven Gables" and other stories. Also (by Fitchburg Railway),

Concord, on the Concord branch of the Merrimac River, noted for beauty of scenery in the neighborhood, and in connection with the literary labors of Henry D. Thoreau, Hawthorne, and others.

ROUTE NO. 5.—EASTERN.

BOSTON TO PORTLAND, QUEBEC AND MONTREAL, BY BOSTON AND MAINE AND GRAND TRUNK ROADS.

Division A.

BOSTON TO AND AT PORTLAND.

Leave Boston by rail on the Boston and Maine railroad, by Medford, Melrose, *South Reading Junction* [connection for North Danvers, Georgetown and *Newburyport*]. Reading, *Wilmington Junction* [connection for LOWELL], and minor stations, to

LAWRENCE, large manufacturing town of the State of Massachusetts, with heavy specialty of cotton and prints, lying on the Merrimac River, whence, by means of a dam, the important water-power is derived. Some of the mills are of immense size and capacity, employing operators to the number of thousands. Has a *Common*, a *City Hall*, and other buildings worth notice; and an *Operative Library* forming a special feature. [Railway connection with LOWELL, and thence with *Nashua* and the North, with Boston and the South, &c.; also Northwest with Manchester, Concord, &c.]

Lawrence by N. Andover, *Bradford* [connection for Georgetown and *Newburyport*], *Haverhill* (with fine long Bridge over the Merrimac to Bradford, and some educational institutions of prominence), Atkinson, Newton, &c., to

EXETER, New Hampshire, lying on Exeter River, and a place of importance in coasting commerce and manufactures; thence by S. Newmarket to

Newmarket Junction [connection west to *Concord* and the Franconia Range of the White Mountains; and east to

PORTSMOUTH, New Hampshire, on the Piscataqua River, second city of the State in importance, with a large and very fine harbor; a *United States Navy Yard;* and a connection, by bridge, with *Kittery,* Maine, also an important naval station. From Portsmouth can be reached *Rye Beach* and *Hampton Beach*, attractive bathing-places on the New Hampshire coast, and the *Isles of Shoals*, off the coast, celebrated fishing and summer resorts.]

Pursuing the main line, Newmarket Junction, by Newmarket, Durham, &c. (within sight of the broad Piscataqua) to

Dover, New Hampshire, thriving town on the Piscataqua. [Connection, west, for Alton Bay, Lake Winnepesaukie, and the White Mountains.]

Dover, by *Salmon Falls* [connection for Great Falls, Rochester, and to *Alton Bay*, &c.]; by *South Berwick Junction* [connection eastward for *Kittery* and *Portsmouth*]; by Wells, Kennebunk, Biddeford, *Saco* (large manufacturing village on the Saco River, with extensive water-power and very handsome Laurel Hill Cemetery, West Scarboro, Cape Elizabeth, and minor stations, to

PORTLAND, commercial metropolis of the State of Maine, and one of the most important cities of the East, lying on a peninsula at the Southwest of Casco Bay, with a very handsome and convenient location and one of the deepest and best harbors on the Atlantic coast. The harbor has many fine islands, and is defended by *Fort Preble* and other extensive fortications. A great fire, in 1866, destroyed a large portion of the city, but the marks are now only visible in the increased beauty of the well-laid-out and handsomely-shaded city. From the *Observatory*, overlooking the harbor, fine views can be caught over the sea and coast, and over the distant country, West to the White Mountains. The most extensive thoroughfare is *Congress street*, which runs the whole distance of the peninsula. Among the most notable buildings are the *City Hall, Court House, Marine Hospital*, and some of the churches, manufacturing and commercial structures. The *Atheneum* and *Mercantile Library* have fine libraries; and the *Natural History Society* possesses an excellent cabinet of varied character. Excursions from Portland include the *Islands in the Bay; Cape Elizabeth*, a favorite bathing and fishing resort on south side of the Bay; *Sebago Pond*; and many of minor interest. Prominent Hotels at Portland: the *United States, Preble, Falmouth, American*, etc.

[The *Allan Lines of steamships* between Liverpool and Glasgow, and Halifax, Quebec and Montreal, make Portland an important depot at all seasons and the port and end of sea-route in the winter season.]

[Railway connection from Portland southward, by route just traversed; to Montreal and Quebec, by Grand Trunk (see route following); to *Augusta, Bangor,* Moosehead Lake, &c., by the Maine Central; to Lake Winnepesaukie and the White Mountains, by the North New Hampshire, &c.]

Division B.

PORTLAND TO QUEBEC OR MONTREAL.

Leave Portland by rail on the Grand Trunk Railway, by minor stations to *Yarmouth Junction* [connection for *Augusta*, capital of the State of Maine, lying on the Kennebec River; for *Bangor*, important town on the Penobscot River; and for towns and sections farther east]; to *Danville Junction* [connection for Bangor, for *Skowhegan* and *Moosehead Lake*]; by minor stations to GORHAM, at the northern edge of the White Mountains [important intersection, by stage-coaches, to and from the mountain towns and resorts; and views of the mountains, in fine weather, peculiarly striking, from all this section of the road]; to *Island Pond* [connection, southwestward, with the Connecticut and Passumpsic River Railway, from Vermont and Franconia Notch sections]; to *Stanstead* [carriage communication with Lake Memphremagog]; to *Sherbrook* [another connection with the Connecticut and Passumpsic

River line]; to *Richmond* [point of divergence of the branches of the Grand Trunk road, to Montreal and Quebec.]

Pursuing the route to Montreal: Richmond by St. Hyacinthe, St. Brune, St. Hilaire, and other unimportant stations, to *St. Lambert*, on the St. Lawrence River, whence ferry to *Montreal*.

Pursuing the route to Quebec: Richmond by Danville, *Arthabasca* [connection, by Bulstrode, to *Doucet's Landing*, on St. Lawrence River, at lower end of Lake St. Peter; thence by ferry to *Three Rivers*, Canada]; by Becancour, Black River, and minor stations, to *Chaudiere Junction* [connection for *Riviere du Loup* and Lower St. Lawrence]; to *Point Levi*, on the St. Lawrence, whence ferry to *Quebec*.

[For notes of **Montreal and Quebec**, see Canadian routes.]

ROUTE NO. 6.—NORTHERN AND EASTERN.

BOSTON TO LAKE WINNIPESAUKIE, THE WHITE MOUNTAINS AND PORTLAND (OPTION OF CANADIAN CITIES)—BY BOSTON AND MAINE RAILROAD, &c.

Leave Boston by Boston and Maine railroad, as by route to Portland, &c. As by that route, to

Dover, New Hampshire. Thence Dover and Winnipesaukie road, by Gonic, *Rochester* [junction with road from Salmon Falls by Great Falls, and its extension northward to *Unionville*] Farmington, Davis', New Durham and Alton, to

Alton Bay, at the extreme southern point of *Lake Winnipesaukie*—pleasant residence, with fine views of the Lake and mountains northward, but deriving its principal importance from the railway and steamer transit through it. From Alton Bay a visit should be paid, if time allows, before proceeding northward, by "Lady of the Lake" or other staunch little steamers on the Lake, to

WOLFBORO', on the eastern side of the Lake, a charming summer resort, with fine views, excellent sailing and fishing, and much attraction and popularity as a residence. Hotel: the *Pavilion*. [Stage-coach may be taken at Wolfboro', for proceeding northward to Conway, without visiting Centre Harbor; but this course is scarcely advisable on a first visit.] Wolfboro', again by steamboat, through charming lake-scenery, to

CENTRE HARBOR, larger village on the northern shore of the Lake, the location of which is considered unequalled by many tourists, as the Lake itself, with its exquisite combination of island groups and wooded shore, with bold mountains forming a background in all northerly directions, is one of the very finest in America and with few superiors elsewhere. Leading Hotel: the *Senter House*. Among the finest points of mountain view, from Centre Harbor and elsewhere on the Lake, may be named *Mounts Salmonbrook, Whiteface, Ossipee, Major, Chicorua, Red Mountain* (ascent by carriage and on horseback), *Kearsarge* and *Monadnock*. Many and charming excursions are made from the village, on the lakes, to the mountains and elsewhere; and among the most notable is that to *Squam Lake*, lying a few miles west of Winnipesaukie very romantic in scenery, and supplying rare trout and other fishing.

At Centre Harbor stage-coach is taken, for remainder of the route northward to the White Mountains. This affords one of the most magnificent rides attainable in the world, especially if fine weather allows outside seats on the coach to be used. The road leads up the Saco River, along the charming *Conway Valley*, with views of the Lake district lingering behind, and others of the great mountain section continually changing ahead, and with *Mount Washington*, the monarch of the eastern range, often in sight. This ride terminates at

NORTH CONWAY, a picturesque village lying in the valley, from which the views of the White Range are something, in comparison, like those of Mt. Blanc from Chamounix, while the number of easy excursions to celebrated points is almost unequalled. It is here that many of the artists' summer sketches, especially of *Kearsarge* and *Chicorua* and the higher peaks of the White Mountains, all in full view, are made; while the *Ledges* (grand perpendicular cliffs, nearly one thousand feet in height) *Artists' Brook*, the *Cathedral, Diana's Bath*, &c., are within convenient reach. Prominent Hotels: the *Kearsarge, Washington, McMillan, Cliff*, etc.

North Conway, by stage-coach or carriage, through *Pinkham Notch*, surrounded by the lesser giants of the White Range, to the

GLEN HOUSE, with much fine scenery in the neighborhood, but especially notable as being the nearest of any of the mountain resorts to the great peaks of the White Range, and giving the rarest views of them—as well as the point from which the ascent of *Mount Washington* is made, by rail. Among the points of interest to be visited from the Glen, are *Thompson's* and *Glen Ellis Falls*, the former on the Peabody River, some two miles from the hotel, and the latter on the Ellis, about four miles; the *Crystal Cascade*, near Glen Ellis; *Garnet* and

Emerald Pools, with peculiar colors indicated by their names, &c. But the speciality of the Glen House, as before noted, is the

Ascent of Mount Washington, by railway. The features of this ascent need no description, especially to those who have made Alpine crossings by rail. It is considered eminently safe, has little fatigue involved, and certainly supplies all the elements of the picturesque and the exciting. Stout clothing is advisable, if not always necessary. This ascent being made in the morning, the top of the giant will be reached at the most favorable hour, and the *wonderful view from the summit enjoyed*, if the capricious weather allows that great privilege. This view is quite equal to that from the Rhigi or Pilatus, over Switzerland, though perhaps lacking the variety in scenery. To the west, in bright weather, are seen the higher peaks of the Green Mountains of Vermont; southwest, some of the White and many of the Franconia Range—especially Lafayette; north and north-east, the other great peaks of the White Range, and more distant the mountains of Canada; east the sea, beyond Portland; southeast and south those surrounding Lake Winnipesaukie, and that Lake itself; while various rivers, small lakes, towns and hamlets combine to make up a picture of marvellous extent and beauty. Dinner is provided at the *Tip-Top House*, on the summit, where during the last seasons scientists have resided all winter, to make observations.

Horses (kept in waiting) and guides should be taken at the summit, and the descent made in the other direction, crossing *Mts. Franklin, Monroe* and *Pleasant;* and the three wondrous gulfs. the *Gulf of Mexico, Tuckerman's Ravine,* and *Oakes' Gulf* (some or all of them containing deep snow in midsummer) will be pointed out by the guides. At the end of nine miles' descent will be reached the

CRAWFORD HOUSE, lying in what is now called the Willey Notch, and nearly at the foot of *Mt. Crawford,* while *Mt. Webster* and other giants of the range show grandly northward. The most marked feature in the neighborhood of the Crawford, is the

Willey House, standing at a short distance up the Notch, where in 1826 a landslide from the mountain above destroyed the family of the same name, and their residence—of which catastrophe many relics are yet pointed out and a few of them still offered for sale. From the Crawford may also be made the

Ascent of Mt. Willard, practicable either by carriage or on foot, and affording a most magnificent series of views from near the summit; as also visit to the *Devil's Den,* a cave of peculiar wildness and some danger of access. Also may be visited, from the Crawford, *Gibbs' Falls,* a cascade of much beauty, reached by short walk from the house.

From the Crawford House, by stage-coach or carriage, by the *White Mountain House, Falls of the Ammonoosuc*, and *Bethlehem*, to the

PROFILE HOUSE, in the Franconia Notch.

[For notes on the Profile House and neighborhood, see termination of route: "New York to the White Mountains, by New London, &c."—Route No. 7.]

[The tourist who has made his arrival at the Franconia Notch by the just completed route from Boston, and who yet wishes to return southward without proceeding to either Quebec or Montreal, should pursue one of the following named routes in return, for the sake of variety in direction and scenery. 1st. From Littleton (stage-coach from the Profile House), by rail by Wells River, Plymouth, Weir's Landing, Concord, Worcester and New London, and steamboat of Norwich and Worcester line from New London to New York. (See route No. 7: "New York to the White Mountains," reversing.) Or, 2d. From Littleton to Wells River, and continue by rail by White River Junction, Bellows Falls, Springfield, Hartford and New Haven to New York. Or, 3d. From Littleton to Wells River, thence on by rail by White River Junction, Rutland, Troy or Albany; and down the Hudson River by boat or rail to New York. Or, 4th. From Littleton to Wells River, to Burlington, boat on Lake Champlain to Ticonderoga (for Lake George) or Whitehall, Saratoga, Albany or Troy, and one of the two last-named routes to New York.]

ROUTE NO. 6.—NORTHERN AND EASTERN.

[To go northward from Littleton to Montreal or Quebec. In either case to White River Junction; thence, for Montreal, by *Burlington* and Rouse's Point to La Prairie; for Quebec, by *Lennoxville* and Arthabasca for Point Levi.]

ROUTE NO. 7.—NORTHERN AND EASTERN.

NEW YORK TO THE WHITE MOUNTAINS AND CANADA, BY NEW LONDON, NORWICH AND WORCESTER, LAKE WINNIPESAUKIE, ETC.; OR BY NEW LONDON AND NORTHERN ROAD.

Division A.

BY NEW LONDON, NORWICH AND WORCESTER, MERRIMAC AND WINNIPESAUKIE ROUTE.

Leave New York at 5 P. M., on Sound, by Norwich and Worcester boat, as by corresponding line for Boston. (See Boston route: "By boat by New London, &c.") As by that route, to *New London, Norwich,* and to

WORCESTER, point of separation of the trains eastward for Boston and northward for the Mountains. Worcester to

Groton Junction [connections eastward for *Concord* and BOSTON; also eastward for *Lowell* and *Lawrence;* also westward for the *Hoosic Tunnel* (immense work of engineering, on the Mt. Cenis plan, not yet completed), for ALBANY, *Troy,* &c.] Very soon after leaving Groton Junction comes into view the *Merrimac River,* with the striking and picturesque scenery of the

Valley of the Merrimac, considered among the

finest rivers in New England, and thenceforward accompanying the traveler almost to the foot of the mountains. Next important point is

MANCHESTER, New Hampshire, large and thriving manufacturing village, on the Merrimac, the mills and some other factories worthy of attention from those whose leisure permits stoppage, but showing even more than ordinary interest in manufacturing detail, even from the train. Hotel: the *Manchester House*. Manchester, still along the Merrimac, to

CONCORD, capital of the State of New Hampshire, lying on the same river, and presenting many points of attraction for visit or residence. It has handsome public grounds; notable public buildings, in the *State House* (recently rebuilt), the *State Lunatic Asylum, State Prison*, and some of the municipal and other erections; and *Main street*, the principal thoroughfare, is remarkably long, fine and well kept. It has great granite quarries in the neighborhood, a considerable amount of manufactures and much general prosperity. Leading Hotel: the *Eagle House*. [Connection by rail, eastward to *Dover, Portsmouth*, &c.; westward to the Connecticut Valley routes northward and southward, &c.] Concord (with distant but very fine views of the Winnipesaukie and White Mountains commencing, and thence continuing, with infinite variations, to the end of the route), by Sanbornton and other stations to

Weir's Landing, at the western edge of *Lake*

Winnipesaukie, with very fine views over the Lake and its many islands. [Connection, by boat on the Lake, for *Centre Harbor* or *Wolfboro'*, and thence by stage-coach for *Conway* and the White Range.] Weir's Landing, by Meredith, &c., to

Plymouth, at the foot of the mountains, with splendid views southward and many attractions as a place of sojourn. Hotel: the *Pemigawasset*. (Leisurely pause, for dinner).

[At Plymouth stage-coach or private carriage may be taken, for the splendid drive of twenty-five to thirty miles, up through the *Valley of the Pemigawasset*, to the *Profile House*—during which a series of views will be enjoyed, approaching the mountains, not often equalled in any land.]

From Plymouth, by rail, literally among the mountains, and among glorious scenery, to

Wells River [connection westward for *Montpelier* (capital of the State of Vermont) and Northern Vermont; southward for *White River Junction*. Route may also be pursued northward, either before or after visiting the Franconia Notch, to *Newport* and the beautiful LAKE MEMPHREMAGOG, lying on the border between Vermont and Canada, and offering fine views, picturesque scenery, excellent fishing, and many other attractions. Or, the same route may be pursued, to Newport, thence on by way of Richmond, &c., to QUEBEC, or bending westward from Richmond, to MONTREAL.]

Wells River, through even grander scenery than that from Plymouth, to

Littleton, small village at the entrance of the Franconia Notch, whence stage-coach is taken, for the ride through the Notch, with many of the best features of American mountain scenery, to the

PROFILE HOUSE. [For notes on the Profile House and neighborhood, see immediately following.]

Division B.

BY NEW LONDON AND THE NEW LONDON NORTHERN ROUTE.

New York by Norwich and Worcester boat on the Sound, 5 P. M., as by route just concluded, to

New London. (Later rest is secured, by this route, than by that by Worcester and Winnipesaukie, from the non-necessity of taking the train until 5 A. M.) From New London by rail, by *Norwich*, through a very pleasant and prosperous part of the State of Connecticut, and past villages embodying large manufacturing interests; by *Willimantic* (manufacturing village, with railway connections westward to the Hartford, Providence and Fishkill road); by Tolland, Stafford, Monson, &c., to *Palmer* [connections eastward to *Worcester* and BOSTON, westward to *Springfield*, for either Hartford, *New Haven* and NEW YORK, or *Pittsfield*, ALBANY and *Troy*]. Palmer to *Amherst*, where the peculiarly splendid scenery of the line, embodying the bold characteristics of the Green Mountain region of Vermont, may be said properly to begin, continuing thence all the way to White River Junction. Amherst to

Grout's Corners, important station. [Connections, eastward to *Fitchburg, Groton* and BOSTON; westward to *Greenfield, North Adams,* &c., and to ALBANY and Troy.] Grout's Corners, by South Vernon, to

BRATTLEBORO, Vermont, thriving town on the Connecticut River, with some manufactures, a specialty of being markedly healthy as a residence, and a *State Lunatic Asylum* bearing a very high reputation. Brattleboro to

Bellows Falls, also on the Connecticut River, deriving its singular name from an ancient peculiarity of the river in the neighborhood. [Connections northwest to *Rutland, Burlington,* Lake Champlain, &c.; and southeast to *Keene, Groton,* BOSTON, &c.] Bellows Falls by Claremont and Windsor, to

White River Junction, at the intersection of the White River with the Connecticut, important place of transfer in cross-travel. [Connections, west to *Rutland, Whitehall, Saratoga,* &c.; northwest to *Montpelier, Burlington, Rouse's Point* and CANADA; east to *Concord, Salem,* BOSTON, &c.] White River Junction to

Wells River. [Northern connections to Newport, Lake *Memphremagog, Quebec* or *Montreal,* as in last previous route.]

Wells River to *Littleton;* thence by stage-coach as before noted, to the

PROFILE HOUSE.

Division C.

AT AND ABOUT THE PROFILE HOUSE (FRANCONIA NOTCH).

Probably no section of mountain scenery in America, of like extent, presents so many points of interest and beauty as appear in the Franconia Notch, though in the detail of absolute grandeur it can by no means claim the same distinction. Taking the Profile House as the central point, the principal objects may be found grouped around it within very brief distance, as follows:

Echo Lake, lying within a few hundreds of yards, embosomed in fine woods, under the brow of *Eagle Cliff*, affording fine views of *Mt. Lafayette*, charming boating, and a repetition of *echoes* (from the Cliff), scarcely second to those of the Eagle's Nest at Killarney. The

Cannon Mountain, at the base of which the Profile House stands, and ascended from it, with moderate difficulty, a magnificent view being the reward. On the top of the mountain, at near the brow overhanging the valley, some rocks, singularly disposed, suggest the shape of a *Cannon*, mounted on its carriage, whence the name; and the extreme brow of the mountain itself forms, in a peculiar combination of great rocks, the

Old Man of the Mountain, colossal face, sixty feet in height from chin to brow, hanging over the immense gulf, and perfect in every detail of a majestic human face, as seen from the road at some distance below the Profile (which of course takes its name from that view). Below this, which is undoubtedly the most striking single curiosity of all the range, lies the little *Profile Lake*, sometimes called the "Old Man's Bath," or "Washbowl," or "Mirror," affording a wonderful reflection of the stony face in calm weather, and said to be full of fine trout. A mile below the Profile is to be reached (ride or walk), the *Basin*, pool of remarkable shape and character, and *Old Man's Foot*, lying in it in colossal stone. Thence, five miles further, the *Flume House*, summer resort, now disused; and near it

The Flume, only second to the Old Man of the Mountain as a great natural curiosity, being an immense fissure or split in the solid rock of the mountain, varying from 10 to 20 feet in width, and the walls from 20 up to 100 feet in height, with a small rapid stream brawling over rough stones below, and a boarded walk up the gorge. At one point, a huge oval stone, of many tons in weight, hangs by the two points midway up the chasm; and at another a dangerous bridge has been thrown over, at the top, by the falling of a tree. At no great distance from the Flume lies

The Pool, a literal hole in the rock, of great depth and singularity of appearance, reached by a difficult climb down the bank, and formerly the abode of a strange madman named Merrill, who paddled visitors round it in a crazy boat, declared it the "centre of the earth," and exhibited a letter to him from Queen Victoria, dated at the Kitchen of Buckingham Palace!

From the front of the Flume House and neighborhood is to be seen a natural wonder of great prominence, the

Dead Washington, being the profile face and form of that hero, in a recumbent position and as if shrouded, lying at a length of miles, the shape supplied by the shapes and position of several mountains of the Haystack group. May also be seen, between the Profile and the Flume, at some distance from the road, *Walker's Falls*, a fine cascade; and two miles below the Flume, *Georgiana Falls*, the largest in the range.

Of Ascents from the Profile, besides that of the Cannon, the principal are those of

Mount Lafayette, the highest peak of the Franconia Range, and commanding a fine view, with only a limited amount of toil (horseback or foot)—and *Bald Mountain*, a lower elevation, but still with fine view (carriage).

[From the Profile House to *Crawford House*, (carriage) for the White Range and ascent of Mount Washington from that direction, with descent by rail to the Glen House; or to *Littleton*, Wells River and White River Junction, for pursuance of the route to Canada.]

ROUTE 8.—NEAR WESTERN.

NEW YORK, BY RAIL, BY THE NEW JERSEY CITIES, TO AND AT PHILADELPHIA.

Division A.

NEW YORK TO PHILADELPHIA, BY THE NEW JERSEY RAILROAD.

Leave New York by the New Jersey Railroad, by ferry from foot of Cortlandt street; cross the Hudson river, to

JERSEY CITY, a large and thriving town, lying in the State of New Jersey, but really a suburb and connection of New York, with which most of its more important business interests are identified. It is growing and improving rapidly, is laying out public grounds, has extensive Water-Works, and enjoys the specialty of the *Cunard Docks* near the ferry, from which sail all the steamers of that popular line. Hotels, *Taylor's, American, Fisk's*, &c. From Jersey City, by rail, across flat and uninteresting country, to

NEWARK, on the Passaic River, now largest city in the State, and one of the handsomest, as well as most important in point of manufactures, especially of leather, carriages and fancy work. [Reached by steamboat and other water-conveyance from New

York. Also reached from New York by train on the Newark and New York road, from foot of Liberty street; and (northern portion) by the Morris and Essex road, from foot Barclay street.] It is regularly laid out; has two handsome parks, many charming drives in the neighborhood; a great number and variety of the residences of the wealthy on *Broad* and other principal streets; some public buildings worthy of attention (including the *Post Office, City Hall, County Court House* and several of the many churches); and is famed for the exceptional beauty of its female population, as seen on promenade or elsewhere. Hotel: the *Newark House*. From Newark visit may be paid to *Orange,* very beautiful village, lying near, at the north; to the *Orange Mountains;* to the popular place of resort, *Llewellyn Park,* &c. Or, they may be reached directly from New York by the Morris and Essex road, foot of Barclay street.] Newark to

ELIZABETH, smaller town somewhat resembling Newark in appearance and general characteristics, though less notable in manufactures and possibly excelling the other in the finished beauty of some of its suburban grounds and wealthy residences. Has the specialty of being passed through, daily, by more railway trains than almost any other town in America, two great lines intersecting in it, and an immense coal-trade from Eastern Pennsylvania passing through it to its adjoining town and the principal entrepot and shipping-port of that article,

Elizabethport. Rivals Newark in female beauty, in fashion and the wealth of residents. [Also reached from New York by the New Jersey Central road, foot of Liberty street.] Elizabeth to

Rahway, handsome village, also much affected as a residence of citizens, and with specialty of considerable manufactures, for the Southern and other markets. Rahway to

NEW BRUNSWICK, one of the oldest towns in the State, though inferior in size to several others. It is pleasantly situated on the Raritan River, has a considerable amount of manufactures, and is the point of entrance into the Raritan River and Bay of the same name (Lower New York Bay), of the *Delaware and Raritan Canal,* from the Delaware River at Bordentown. Its principal celebrity, however, lies in its being the seat of *Rutgers College,* and the *Theological Seminary* of the Reformed Dutch Church, both old and influential institutions, holding excellent rank. Among the buildings best worth notice, are those of the *College,* on an elevated square, within view from the railway; the *Theological Hall;* the *County Buildings,* in the public square; and several churches of prominence. Many fine drives are to be enjoyed by those making stay, into the handsome and well-cultivated country in the neighborhood, to *Bound Brook,* &c. Leading Hotels, the *Railroad, Bulls Head,* &c.

Beyond New Brunswick, the railway is accompanied for much of the distance by the Delaware and Raritan Canal. At *Monmouth Junction* intersection is made for *Freehold*, the *Battle Ground of Monmouth*, and south-eastern portions of the State. Next stopping place of importance,

PRINCETON—Station, the town lying away at the right, though in sight, and steam connection in waiting. Princeton is another of the old and important towns of the State, with a peculiar status in Revolutionary history as having been the scene of one of Washington's most memorable conflicts, of which the field extends from the town itself to what is called the "Battle Ground," more than a mile distant. Its more marked celebrity, however, lies in its being the seat of the *College of New Jersey* (called alternately, "Princeton College," and often, from one of the oldest buildings, "Nassau Hall.") It is also the seat of the *Theological Seminary* of the Presbyterian denomination, in connection with the College, which has long enjoyed a very high reputation, and which is now presided over by the celebrated Scotch divine, Dr. McCosh, late of Queen's College, Belfast. The grounds of the College (like many of those of the town) are very handsome. Peale's "Washington," in the College library, is a picture of merit and historical interest. To the College has also lately been added an *Astronomical Observatory*, with fine instruments.

Beyond Princeton is soon reached

TRENTON, capital of the State of New Jersey, lying on the left or east bank of the Delaware river, and famous as the scene of Washington's "Crossing the Delaware," January, 1777. It is a thriving manufacturing town, especially preëminent in iron works; and has very costly constructions connected with the passage through the town of the Delaware and Raritan Canal. [Railroad connection, north, for *Belvidere, Easton,* and the Upper Delaware and Pennsylvania Coal-Regions; and south to *Bordentown,* and by that route to PHILADELPHIA.] The principal erections of prominence are the *State House,* modern and very handsome (with valuable Revolutionary memorials in the Library); the *State Lunatic Asylum, Arsenal, Penitentiary,* and some of the *County buildings.* The views over the Delaware and the Pennsylvania shore opposite, from some portions of the town, are very fine and memorable. Prominent hotels, the *American* and *Trenton.*

[Optional route may be taken, at Trenton, by rail, down the Delaware River to *Bordentown* and *Camden,* thence to PHILADELPHIA by ferry. See Division B. of this route.]

At Trenton the Delaware River is crossed, by bridge, to the State of Pennsylvania, by

Bristol, handsome and thriving village of that State, and by *Frankford* (with a United States Arsenal) to *Kensington* and *West Philadelphia,* point of debarkation for

PHILADELPHIA.

Division B.

NEW YORK TO PHILADELPHIA, BY BOAT AND RAIL OF CAMDEN AND AMBOY LINE.

Leave New York by Camden and Amboy boat, from Pier No. 1 North River (Battery), down the Bay of New York to the entrance of the Great Kills or Staten Island Sound, thence up that Sound, with Staten Island on the left and the New Jersey shore on the right. Only places of consequence passed, on either side, *Bergen Point*, on the right, favorite place of summer resort; also on the right, *Elizabethport*, with extensive coal-wharves and small shipping; and, also on the right, *Perth Amboy*, old but decayed seaport, once expected to rival New York—to

South Amboy, New Jersey, end of the route by boat and commencement of rail.

(Or, according to weather, down the Bay of New York to and through the Narrows, and up the Lower Bay, with Staten Island at the right and the Quarantine and the distant shores of New Jersey on the left, making only the landing at Perth Amboy, to *South Amboy*, place of disembarkation as before).

South Amboy, by rail, by *Washington*, and *Spotswood*, to

Jamesburg. [Connection southward to *Freehold*, thence to Long Branch and the south-east; and northward to the New Jersey Railroad at *Monmouth Junction*, for Newark, Trenton, &c.] Thence to

Cranberry, Hightstown [connection to *Pemberton* and southwest] and

BORDENTOWN, handsome large village on the Delaware, and point of entrance into that river of the Delaware and Raritan Canal; famous as having long been the residence (at Point Breeze—grounds to the right) of Joseph Bonaparte, ex-king of Spain. [Connection by boat down the Delaware to PHILADELPHIA; by rail to *Trenton*, &c.] Bordentown to

BURLINGTON, also lying on the Delaware, and considered one of the handsomest towns in the State. It is the seat of *Burlington College* (Episcopal), and of several notable male and female schools. [Connection by boat to PHILADELPHIA and *Bordentown;* by rail to *Mount Holly* and other towns in the interior of the State.] Burlington by *Beverley* and other minor places, to

Camden, on the Delaware, opposite Philadelphia. Thriving town, with much agricultural and some manufacturing industry, and residence of many Philadelphians. [Connection by West Jersey Railroad southward to *Bridgeton*; south-eastward to *Millville* and CAPE MAY, favorite sea-coast resort at the Capes of the Delaware; eastward by the Camden and Atlantic to *Atlantic City,* another favorite watering place on the New Jersey coast of the Atlantic; and by Pemb. and Hightstown road to the New Jersey Southern, *Long Branch* and New York]. Ferry across the Delaware to

PHILADELPHIA.

Divison C.

AT AND ABOUT PHILADELPHIA, WITH EXCURSIONS.

PHILADELPHIA, most important city of Pennsylvania, second in the Union in point of population, largest of all in the extent of ground comprised within city limits, and dividing with Boston the claim of being the most influential after the commercial metropolis—lies on the Delaware River, at about one hundred miles from its mouth at Delaware Bay, and above and very near the debouchure into that river of the Schuylkill, the latter stream running through the city at its western extremity and adding materially to the beauty and healthfulness of location. It is well known to have been founded by William Penn, the Quaker, and to be the headquarters of his denomination (whence its *soubriquet*, the "Quaker City"); and it enjoys, in addition, the distinction of being the most regularly built city on the continent if not in the world, the mass of its streets lying at right angles and giving it an appearance of primness alternately counted a charm and a blemish. From this latter feature, combined with all the streets running parallel with the Delaware being numbered, from one upward—and the space between each of these streets, on the intersecting ones, numbered as one hundred, in supplying street-numbers—less difficulty is involved, in the stranger finding his way through and about it, than through any other city in the world, of corresponding size. Still additionally it should be noted that the street-car system is wonderfully complete and perfect, routes crossing each other at short distances, and a system of "transfers" from one route to another making transit much easier and cheaper than it could otherwise be found. Carriage-hire, cheaper than in New York, though high; not differing materially from the same detail at Boston.

Philadelphia has many notable Streets, of which the characteristics are worth study, for their individual and collective character. First among these is *Chestnut Street*, at once business and fashionable, on which are located some of the best hotels, and which has by far the handsomest display of shop-fronts on the continent. Next to this, perhaps, is *Market Street*, wide thoroughfare, dividing the cross-streets into "North" and "South," and displaying much railway traffic and other heavy trade. *Arch* and *Walnut* are also both business streets of importance. The Exchange stands in *Dock Street*, between Walnut and Spruce; and much of the commercial and financial force of the city is to be found in that neighborhood, and near the Delaware, between *Shippen Street*, on the South, and *Vine Street*, on the North, and *Front Street* to *Sixth Street*, in the cross direction. *Fourteenth Street* is ordinarily called *Broad Street*, and has much fashion and many prominent buildings. *Ridge* and

Girard Avenues hold position as places of fashionable residence, and drives leading to Girard College, Fairmount Park, etc.

Of Public Buildings there are many of importance and interest. The first place is held, historically, by *Independence Hall*, Chestnut street, notable as having been the place of signing of the Declaration of Independence from Great Britain, Fourth of July, 1776. Some historical pictures of value, statues, and many relics are preserved there; and among others the "Liberty Bell," rung at the time of the Declaration, and bearing the strangely appropriate inscription: "Proclaim liberty throughout the land to all the inhabitants thereof." (Admission to the Hall, every day, 9 to 2). The building and wings are now used as public offices. A recently erected statue of Washington fronts the main entrance. Next of the public buildings in importance, is *Girard College*, on Ridge Avenue, some two miles from the city centre—the several buildings modern and of fine architecture, and the grounds handsome, but its principal celebrity (it has very little as an educational institution) lying in the strange will and bequest of Stephen Girard, the merchant, which founded it, and which among other odd features, allows no clergyman to enter it even on a visit. Next in importance is the *United States Mint*, Chestnut street, with very perfect and interesting processes and a splendid collection of coins (admission daily, 9 to 12). Besides these, there are the *Custom House* (formerly the United States Bank), Chestnut street; the *Exchange*, Dock street; the *University of Pennsylvania*, Ninth street near Chestnut; *Jefferson Medical College*, Tenth street near Chestnut; the *Pennsylvania Hospital*, Pine street; *Pennsylvania Insane Asylum*, West Philadelphia (with West's great picture of "Christ Healing the Sick"); *U. S. Marine Hospital*, near the Navy Yard; the *Franklin Atheneum*, and other library and literary buildings; *Pennsylvania Academy of Fine Arts* (with many good pictures: open daily), Chestnut street; *Eastern Penitentiary*, Coates street, near Girard College; *Union League Club House*, Broad street; *Masonic Temple*, Broad street; *Ledger Building*, corner of Sixth and Chestnut streets, etc.

In Antiquities Philadelphia possesses, besides *Inpendence Hall* (already mentioned), *Carpenter's Hall*, Chestnut street, used for the first assembling of the Colonial Congress; *Hultzheimer's*, where Jefferson wrote the Declaration, cor. Market and Seventh streets; the *Grave of Franklin*, cor. Arch and Fifth streets; *Indian Queen Hotel*, once residence of Jefferson, cor. Market and Front streets; the *Old Penn House*, near Fairmount; part of *Penn's Elm Tree*, in collection of Historical Library Association; and others of minor importance.

Of Public Grounds Philadelphia has more than the average in both variety and beauty. FAIRMOUNT PARK, on the Schuylkill (in connection with the long-celebrated *Fairmount Water Works*), is one of the largest parks in the world, and has much beauty in grounds and views, though little more than commenced; and a bronze sitting statue of Lincoln has recently been inaugurated at near the Schuylkill entrance, while cheap service-carriages and all conveniences to visitors are supplied. The finest view is from *George's Hill*, and the finest drive, *Vista Drive*. The *Water Works* themselves demand attention, as among the best of their class; the views over the Schuylkill from the raised promenade are notably fine; and the *Suspension Bridge*, at the same point, is the most interesting structure of that character at or near the city, it having been built by Col. Ellett, the constructor of the Niagara Suspension Bridge, and afforded a model for the latter. (Other Bridges of interest are the *Iron Bridge*, over the Schuylkill at Chestnut street; the *Market Street* Bridge, of wood, very old; &c.) (There are also other Water Works: the *Delaware*, on the river, foot of Wood street, and the *Western*, with a beautiful tower, opposite Fairmount.) Of the other public grounds of the city, the most interesting are *Independence Square*, rear of Independence Hall; *Washington Square*, near it; *Logan Square* (largest of the old), Eighteenth street; *Franklin Square*, Race and Sixth streets; *Penn Square*, Broad and Market streets; *Jefferson* and *Rittenhouse Squares;* and *Hunting Park* (old race-course) on the York road.

Among the most notable of Philadelphia churches, are the *Cathedral of St. Peter and St. Paul* (Catholic), Logan Square, with a noble dome, an admired altar-piece, and some good paintings; *St. Mark's* (Epis.), Locust street, with tower and spire of peculiar beauty; *St. Paul's* (Epis.), Third street; *Christ Church* (old), Second street, with tall steeple, fine chime of bells, and communion service of the time of Queen Anne; *Church of the Incarnation*, Broad street; *Baptist*, Broad street; *Calvary* (Pres.), Locust street; *St. Stephen's* (Epis.), Fourth street; *St. Peter's* (old), Pine street; *St. Andrew's*, Eighth street; &c., and (as curiosities, though eschewing any attempt at architecture) many of the *Friends' or Quaker Meeting Houses*, of which the city has a remarkable number and variety.

Of Libraries, there are a large number, though the aggregate of volumes embraced in all does not reach far beyond a quarter million. Among them are the *Franklin* (sometimes called the "Philadelphia,") South Fifth near Chestnut street; the *Atheneum*, Sixth street; the *Mercantile;* the *Apprentices'*, *Friends'*, *Law Association*, &c.; besides those connected with those prominent institutions, the *Historical Society*, Sixth and Adelphi streets (antiquities and curiosities); *Academy of Natural Sciences*, Broad street; the *Franklin Institute*, Seventh street, &c. The principal Art Gallery is the *Pennsylvania Academy of Fine Arts*, Chestnut street, containing among other prominent pictures, West's

"Death on the Pale Horse," Allston's "Raising of Lazarus," and others of merit by Stuart, Sully, Leslie and others. The principal Market, and one of the best-arranged and most luxuriously-supplied in America, is located on Market street, in the lower part of the city, and will well repay a visit, for observance of the varied productions of the surrounding country.

Principal Places of Amusement: the *American Academy of Music*, Broad street, the handsomest and one of the largest musical houses in the United States; *Arch Street Theatre*, street of the same name; *Chestnut street*, street of that name; *Walnut street*, street of same name; *American*, Walnut street; *Carncross and Dixey's Opera House* (Ethiopian), Eleventh street; *American Museum*, Ninth, and Arch streets, &c. Leading Hotels: the *Continental*, Chestnut street; *La Pierre House*, Broad street; *Colonnade*, Chestnut street; *Girard House*, Chestnut street; *American*, Chestnut street; *St. Cloud*, Arch street; *Washington*, Chestnut street; *Merchants'*, Fourth street.

Surburban and other Excursions of interest, include the

United States Navy Yard, on Front street and the Delaware River, entrance from foot of Federal street; with immense Sectional Dock, stocks and materials for war-vessels, munitions of war, &c. [Walk, or street-car.] Arrangements have been made for the occupation of *League Island*, lower down the Delaware, as a new and larger navy yard, for the laying up of vessels in ordinary; but they have not yet been carried into effect. Of scarcely less interest are the

U. S. Arsenals, of which one of the most important is to be reached at *Frankford*, north-east of the city, with interesting collection of arms and the largest powder magazine in the country; and the other near *Gray's Ferry*, south of the city. Also,

Laurel Hill Cemetery, on Ridge Avenue, near the Schuylkill, and considered one of the handsomest of the cemeteries of the great cities, on account of height of location, fine river-view, tasteful monuments and adornments. The group of "Old Mortality," by Thom, at the entrance, and the Chapel, deserve attention, as do many of the monuments to well-known men, among others those of Dr. Kane, Gen'l Mercer, Gen'l Patterson, Dr. Bird (the novelist), Joseph C. Neal, Charles Thompson, Hassler, &c. [Reached by street car, drive, or boat up the Schuylkill from Fairmount.] Second in importance are the *Woodlands Cemetery*, on the Darby Road, west of the Schuylkill; *Monument Cemetery*, Broad street; *Glenwood Cemetery*, Ridge Road; *Mount Vernon Cemetery*, Ridge avenue; *Ronaldson's Cemetery*, Shippen street; *Friends' Burial Ground*, Arch and Fourth streets, &c. [All, beyond short walk from leading hotels, reached by street-car.]

Other Excursions, to

The Wissahickon, creek or small river of marked

shaded beauty, emptying into the Schuylkill. [Drive, on Ridge avenue, past Laurel Hill, or trip by boat on the Schuylkill from Fairmount, in the course of which may also be seen the *Falls of the Schuylkill*.] To the *Old Bartram Mansion*, with Revolutionary reminiscences and a Botanic Garden, on the West bank of the Schuylkill. [Street cars on Darby road.] To *Penn's Rock*, on the Haddington road (stone said to have been raised by William Penn).

To *Germantown*, site of the Battle of that name, fought by Washington in 1777; with interesting reminiscences, in Chews' House, the Headquarters, Buttonball Tree Tavern, &c. To *Manayunk*, on the Schuylkill, with water-power and heavy manufactures. [Street cars on Ridge-road, or boat on the Schuylkill.] [Street car and short steam connection, every quarter-hour.] To *Greenwich Point* and *Gloucester Point*, on the Delaware, favorite near places of summer resort, a few miles below the city. [Ferry from South street.] To *Red Bank* and *Fort Mifflin*, two miles below the places last named, with Revolutionary reminiscences, Count Donop's Grave, a Battle Monument, &c.; and also to *League Island*, lying near, and the site of the new Navy Yard. [Boats, very frequent.] To *Smith's Island* (Windmill Island), lying in the Delaware, midway between the city and Camden, and passed through by the ferry-boats. Resort for relaxation and "clam-chowders." To *Camden*, New Jersey [several ferries: see route from New York, Division B.] To *Bridgeton*,

New Jersey, great fruit-packing centre. [Ferry to Camden, and West Jersey Railroad.] To *Vineland*, New Jersey, great grape and fruit growing centre. [Ferry to Camden, and Camden and Atlantic road to Atsion—thence Vineland Railway. To *Bordentown* and *Burlington*. [Boat on the Delaware, or rail.]

Longer Excursions will be those to

NORRISTOWN, on the Schuylkill, county seat of Montgomery County, with pleasant location, two fine Bridges, and handsome Court-House. [Railway on Reading road, or long drive of much beauty]. To

EASTON, DELAWARE WATER-GAP, &c. [See Longer Excursions from New York.] [Rail, on Northern Pennsylvania, and Delaware, Lackawanna and Western roads.] To

HARRISBURG, Capital of the State of Pennsylvania, by *Lancaster*, &c. [Rail on the Pennsylvania Central Road: see routes following.] To

ATLANTIC CITY, favorite place of summer resort, with fine bathing, on the New Jersey coast, near Egg Harbor and the Inlet of the same name. Prominent Hotels, the *Atlantic House*, and *Surf House*. [Reached by ferry to Camden, thence rail on the Camden and Atlantic road, direct.] To

CAPE MAY (Cape Island), still more prominent and popular as a place of sea-side summer resort, and especially chosen by Philadelphians. It lies at the extreme southern point of New Jersey, at the northern

ROUTE NO. 8.—NEAR WESTERN.

entrance of Delaware Bay, has an extensive beach with fine sea-view and bathing, and ranks beside Newport and Long Branch. Prominent Hotels: the *Stockton House, Congress Hall, United States, West Jersey, Columbia, Delaware, Atlantic*, &c. [Reached by ferry to Camden, thence by rail on the West Jersey, and Millville and Cape May roads.] To

LONG BRANCH. [See Longer Excursions from New York.] [Reached by ferry to Camden, thence rail on Pemb. and Hightstown and New Jersey Southern roads. Also, with connection, Long Branch to NEW YORK.

ROUTE NO. 9.—WESTERN AND SOUTHERN.

PHILADELPHIA, BY WILMINGTON (DEL.) TO AND AT WASHINGTON AND RICHMOND.

Division A.

PHILADELPHIA TO BALTIMORE BY WILMINGTON.

Leave Philadelphia by rail on the Philadelphia, Wilmington and Baltimore road, from West Philadelphia. First point of interest passed is the *Lazaretto*, on the bank of the Delaware, some ten miles below the city—an immense building, with cupola, long used for the detention of cases of infectious disease. In a short distance is reached

Lamokin Junction [with the Philadelphia and Baltimore Central Railroad, for *Port Deposit, Havre de Grace*, and Baltimore direct, avoiding Chester and Wilmington.] Beyond Lamokin, continuing by P. W. and B. road, is reached

CHESTER, the oldest town in the State and at one time, under William Penn, the seat of government of the province. It has, as curiosities, the spot where Penn landed on his first coming from England, a very old *Court House*, &c. Very little beyond, the crossing is made from the State of Pennsylvania into that of Delaware; and still a little beyond is passed the *Brandywine Creek*, scene of the battle of the same name (at Chadd's Ford), defeat of the Americans and wounding of Lafayette, in 1778. After several minor stations, is reached

WILMINGTON, Delaware, one of the most important towns of that small State, and in the midst of an agricultural section of special fertility, the great peach-growing district being within easy reach of any one making brief stoppage. It occupies the site of the old Swedish Fort Christina; has extensive shipyards, flour and powder-mills, foundries, &c.; and is also distinguished as the seat of *St. Mary's College* (Catholic), and other educational institutions of merit. Among its most prominent curiosities are the ship-yards and powder-mills, before named; the *Old Swedes' Church*, nearly 200 years old, with ancient grave-yard and singular epitaphs; the College, &c. [Railway connection south to *Elkton, Townsend,* DOVER (capital of the State), *Lewes, Salisbury, Crisfield* (for boat to *Norfolk*) &c,; westward to *Hanover,* HARRISBURG, &c.] From Wilmington, passing *New Castle Junction* [connection for *New Castle,* &c.], and minor stations, is reached

Havre de Grace, Maryland, at the debouchure of the Susquehanna River into Chesapeake Bay, and also at the southern terminus of the Tidewater Canal. Here the Susquehanna is crossed by a handsome and costly Railroad Bridge, not long finished; and in crossing, splendid views are caught (below) of Chesapeake Bay and the shore-scenery on both sides.

[Railway connections from Havre de Grace, northwestward, to HARRISBURG and the West and Northwest.] From Havre de Grace, over flat and low country, with passage of the long

Bridges over Bush and Gunpowder Rivers (the former 5-8 of a mile in length, and the latter 1 mile), both of which were destroyed during the secession-war, and rebuilt,—to BALTIMORE.

Division B.

AT AND ABOUT BALTIMORE, WITH EXCURSIONS.

BALTIMORE, on the Patapsco River, branch of Chesapeake Bay, most important town in the State of Maryland, seaport of eminence, considered one of the handsomest cities in the Union, and dividing with two or three others the claim of producing the most beautiful women, while to Europeans it possesses the peculiar interest of having supplied wives to a remarkable number of the English aristocracy (Wellesley family, and others), and also a wife (Miss Patterson) to Jerome Bonaparte. It has a striking situation, on rising ground sloping up from the harbor, in that respect rivalling Boston; and the numerous spires and monuments fitly crown a picture otherwise of great beauty. Baltimore has an inner and outer harbor, above and below *Fell's Point*, into the latter of which the largest ships enter without difficulty; and the city proper is divided, nearly North and South, by a narrow stream with many bridges, called *Jones' Run*. A strong and handsome fortification, *Fort McHenry*, defends the harbor, and figured conspicuously in both the war of 1812 and that of the secession. Among the chief boasts of the city, and the first objects of interest to the traveller, are

The Monuments, so notable that they have given to Baltimore the soubriquet of the "Monumental City." The first in importance is the *Washington*, in an elevated position on Mt. Vernon Place, at Charles and Monument streets—a base and shaft reaching 200 feet in height, with a statue surmounting all, of "Washington Resigning his Commission." (Accessible, and fine view from balcony at top.) Next in interest is the *Battle*, at Calvert and Fayette streets—a Roman column, with emblematical sculptures, in honor of those who fell in defence of the city, in September 1814. The third, or *Armistead*, in honor of the defender of Fort McHenry in 1814, is merely a tablet, on North Calvert street, and only of interest in the patriotic connection.

Of streets, the most important is *Baltimore street*, running east and west the whole length of the city, and really its Broadway or Regent street. *Holliday, Calvert, Fayette, Lexington, Eutaw, Madison, Park, Saratoga, North Charles, Mt. Vernon Place, Charles avenue*, and other streets on the west side of Jones' Falls, are among the notable; and as centres of business, *Lombard, Caroline, Bank, Gay, High, Market, Broadway*, and other streets on the east side, with those surrounding the City Dock (basin) and principal wharves, lying in that vicinity. Of Public Buildings, among the most notable are the *Exchange*, Gay street, with noble dome; (*Custom House* and *Post Office* occupying part of the same building); the *Maryland Institute*, Baltimore street, devoted to industrial exhibitions, fairs, &c., and a Market; the *City Hall*, Holliday street; *County Court House*, Monument square; *U. S. Court House*, North and Fayette streets; *Penitentiary and Prisons*, Madison street; *Corn Exchange*, South street; the *Shot Tower*, Front and Fayette streets; Of Churches, in Baltimore as in Philadelphia, the most imposing is the Catholic, the *Cathedral*, at Cathedral and Mulberry streets, being the finest ecclesiastical edifice in the city, with impressive towers and dome; one of the largest organs in the country; and two pictures of great value within, a "Descent from the Cross" and "St. Louis Burying His Dead," respectively the gifts of the French Kings Louis XVI. and Charles X. After this, in architectural interest, come the *Unitarian*, North Charles and Franklin streets; the *Presbyterian*, Madison and Park streets; *Grace* (Epis.), Monument and Park streets; *St. Paul's*, Charles street, and many others, the city being by no means deficient in this detail.

Of Literary Institutions and their edifices, may be named the *University of Maryland*, with celebrated Medical Department, Green and Lombard streets;

the *Peabody Institute* (founded by the late George Peabody), Charles and Monument streets; *St. Mary's College* (Catholic), Franklin and Greene streets; *Maryland Historical Society, Baltimore Library, Mercantile Library*, &c., rooms in the *Atheneum*, Saratoga and St. Paul streets; *College of Loyola* (Catholic), Madison and Calvert streets; *College of Pharmacy*, North Calvert street; &c. Principal Theatres: the *Holliday Street*, street of same name; the *Front Street*, or *American*, Front street; *Baltimore Museum*, Broad and Calvert streets; *Grand Opera House* (new); *Concordia* (German), South Eutaw street. Prominent Hotels: *Barnum's*, Monument square; the *Eutaw*, W. Baltimore street; *Gilmour's*, Baltimore street; the *Fountain*, Light street; the *Maltby*, Pratt street.

Cemeteries of prominence: *Green Mount*, Belvidere street and York avenue, with fine gateways and many handsome walks and monuments; *Loudoun Park*, also with fine gateway, Frederick road; *Baltimore Cemetery*, North Gay street; *Mount Olivet*, Frederick road; *Mount Carmel, Western*, and other minor. Other Parks and Public Grounds: *Druid Hill Park*, very large and handsome grounds, recently laid out, in the Northern suburbs [street-car from city centres]; *Patterson Park*, East Baltimore street, with remains of earthworks of war of 1812; *City Spring Grounds*, North Calvert street; *Union Square*, West Lombard street; *Federal Hill*, with Signal

House and one of the very best views of the city and harbor; *Franklin Square*, Fayette street; *Jackson Square*, Hampstead street; etc.

Favorite Excursions, among others, to

Fort McHenry and *North Point*, entrance of the harbor (before spoken of); to *Franklin*, the Convent, &c., by the Frederick road [favorite drive]; to *Govanstown*, by the York road [drive]; to *Catonsville* and *Ellicott's Mills* [horse-car]; to *Towsontown* (military barracks, &c.) [horse-car]; to *Bel-Air*, *Franklinton*, &c.] [stage-coach]. *Down the Chesapeake Bay* [boat, very frequent from harbor-wharves, during the warm season]. Longer Excursions, among others, to

ANNAPOLIS, Capital of the State of Maryland, and seat of the celebrated national *Naval Academy*. It lies on the little River Severn, near Chesapeake Bay; has a history of interest, dating back to 1649; was the spot where Genl. Washington resigned his commission at the close of the War of Independence; and has, in addition to the other attractions named, an educational institution of prominence, *St. John's College*, a *State House*, and much fine river and coast scenery in the neighborhood. [Reached from Baltimore by the Baltimore and Washington road to *Annapolis Junction*, thence branch road direct.] Also, to

NORFOLK, Virginia, on the Elizabeth River, at the extreme southern point of Chesapeake Bay, and the second town in Virginia in point of population.

It has a fine harbor, with great depth of water; and is one of the greatest markets of wild-fowl (especially the celebrated "canvas-back" ducks of the Chesapeake), oysters, fruits and other supplies, to be found south of Philadelphia. Across the river from it are the *Portsmouth Naval Depot*, formerly the most extensive in the Union, but materially damaged by fire at the commencement of the secession war (1861), with the burning of the Pennsylvania, Merrimac and other war vessels,—and the *Gosport Navy Yard*, with Dry Docks of great size and cost. Norfolk and Portsmouth harbor proper are defended by *Fort Calhoun* and the works on *Craney Island;* while the entrance from the sea is commanded by *Fortress Monroe*, the largest fortification in the United States, erected at *Old Point Comfort* (also place of favorite summer resort), on the opposite or north side of the wide mouth of the *James River* (Hampton Roads), directly north and some fourteen miles distant. In the latter neighborhood may also be visited, *Elizabeth City* and *Hampton,* more or less interestingly connected alike with the early history of Virginia and the secession troubles of 1861-5. May also be visited, from Norfolk or Old Point Comfort [boat], the site and *Ruins of Jamestown*, place of first settlement in Virginia, and scene of the romantic episode of Captain John Smith and Pocahontas. [From Baltimore to Richmond by boat, every day during summer, with fine views throughout, and including a distant one of the great dome of the Capitol at Washington.] [Norfolk to RICHMOND, by boat up the James River; or may be visited from Richmond, by same conveyance.] Also, to

Point of Rocks, splendid pass of the Potomac River between Maryland and Virginia, by the *Thomas Viaduct, Ellicott's Mills*, the *Oliver Viaduct*, the *Tarpeian Rock, Monocacy* [branch road to *Frederick*], through the *Monocacy Valley* and other markedly fine scenery; and to

HARPER'S FERRY, on the Upper Potomac, at the intersection with that stream of the Shenandoah, with mountain and river scenery of the first magnificence, and the additional interest of having been the site of an important U. S. Armory and Arsenal (destroyed by fire in April, 1861), and the spot where John Brown, of Ossawatomie, made his celebrated raid and virtually commenced the conflict of the secession, in October, 1859. *Maryland, Bolivar* and *Loudon Heights*, and their fortifications, demand notice, as do a thousand natural beauties and warlike reminiscences certain to be suggested on the spot [Reached from Baltimore by Baltimore and Ohio Railroad.] [Route may be pursued from Harper's Ferry to *Cumberland*, PITTSBURG and the West, without return to Baltimore.]

Division C.

BALTIMORE TO WASHINGTON.

Leave Baltimore by train on the Washington Branch of the Baltimore and Ohio Railroad, from Camden Station, by *Washington Junction* [whence connection, by the main line of the same road, westward to *Harper's Ferry, &c.*]; and by Hanover, Dorsey and other stations, to

Annapolis Junction [connection to ANNAPOLIS, as see "Excursions from Baltimore," preceding]; thence by Savage, Laurel, White Oak Bottom, and other stations, to

Bladensburg, small town on the Eastern Branch of the Potomac, with a mineral-spring and some celebrity as a healthful summer resort for Washington residents and others near; but much more as the scene of the disgraceful defeat of the American by the British forces, in August 1814, immediately previous to the temporary occupation of the Capital— and also as the spot long famous as a duelling-ground for Congressional and other disputants. Very soon after leaving Bladensburg is caught, what should by no means be lost, the

First View of the Washington Capitol, scarcely second to the corresponding distant view of the dome of St. Peter's, in approaching Rome—the dome of the Capitol, since re-erection, being among the largest and finest in the world, and the first sight peculiarly impressive. But a little time and distance, following, before entering the somewhat straggling city, and disembarking at the foot of Capitol Hill—WASHINGTON.

Division D.

AT AND ABOUT WASHINGTON, WITH EXCURSIONS.

WASHINGTON, capital city of the United States of America, and on many accounts specially interesting, as bearing the name of the Great Soldier and Patriot, as having been selected by him as the seat of Government, as having been the scene of all the central legislation of the country and many of its other historical events, and as possessing a location with many marked advantages and certain equally marked disadvantages almost counterbalancing the favorable,—lies in the District of Columbia, on the north bank of the Potomac River, at some 70 miles from the embouchure of that river into Chesapeake Bay, and about 30 miles directly westward from the nearest shore of that Bay, at a little southward of Annapolis. It supplies a geographical meridian of importance: Lat. 38° 53′ 39″ N.; Long. 77° 2′ 48″ from Greenwich; and lies in a direct line, about 120 miles south-west from Philadelphia, and about 200 in a corresponding direction from New York. It dates, as the Capital, from the removal from Philadelphia, about 1800, though the corner-stone of the Capitol was laid in 1793. The city, as a municipality, dates from nearly the same time—is large in extent, and by no means compact in its character; that fact having given rise, many years ago, to the irreverent soubriquet: the "City of Magnificent Distances."

Among the undeniable advantages of Washington, before referred to, are its picturesque location, with elevations and fine views at two points,—those of the Capitol, at the south-eastern end of (main) Pennsylvania Avenue, and the President's House and principal Government Buildings, at the north-western end; its ease of access from the sea, and accessibility by railway from various important points; its moderate climate in winter, principal season of legislative assemblage; and its proximate centrality, as between North and South. Among the disadvantages may be named the doubtful healthiness of some portions (including the Executive Mansion) in summer; and its entire want of centrality towards the limited East and the widely-extended West—the latter feature having given rise to much dissatisfaction, of late years, and some efforts at effecting the removal of the seat of government to some one of the Western Cities—St. Louis being oftenest named. The governmental conveniences now existing on this spot, however, and the late completion of the enlarged Capitol, render it entirely improbable that any such removal will take place within the present century, and leave Washington to be visited and considered as the permanent capital of the United States.

Of course first among its attractions, to the tourist, at whatever season, will be found

THE CAPITOL, standing on Capitol Hill, fronting east and west, and occupying the same site as the original, founded by Washington and burned by the British in 1814, when the Congressional Library, many valuable pictures, the President's House and other buildings, shared the same fate. The present structure, undeniably one of the noblest government buildings in the world, and with many grand and beautiful details in architecture, is an enlargement of that which replaced the burned building, and has consequently the blemish of showing some incongruity in materials and "afterthought" in design. It is immense in extent, however, the entire length being some 750 feet, with a wing depth of 300 and a body depth of 200; and the whole space of ground covered is said to be three and a half acres. Handsome grounds surround the Capitol, with fine shade trees and some good landscape gardening; and from these grounds, below either front, and from the two fronts themselves, remarkably fine views may be obtained.

The *East Front*, (main) has an immense colonnade and portico, with Persico's statues of Columbus, of Washington, allegorical figures of Peace and War, Greenough's "Civilization," etc., on the portico and in the grounds adjoining; and it is here that the Inauguration Ceremonies of each incoming President take place, the auditory filling the portico and the grounds below. The *West Front*, less elaborate, has the view down Pennsylvania Avenue and over the city. The next most prominent feature, and perhaps the most notable of all, is

The Dome, before spoken of as among the noblest in the world. It surmounts the centre of the pile, rising to a height of nearly 400 feet, crowned with a colossal statue of Freedom, by Crawford; and is ascended, from within, by a spiral stairway, for the extensive and magnificent view of Washington, the Potomac, the near portions of the District, of Virginia, Maryland, etc. In the *Rotunda*, immediately under the dome, are to be seen the eight large historical pictures, "Discovery of the Mississippi," "Baptism of Pocahontas," "Declaration of Independence," "Surrender of Burgoyne," "Surrender of Cornwallis," "Washington Resigning his Commission," and "Landing of Columbus." The Rotunda has also a "Massacre of the Innocents," portraits of Lincoln and others, some interesting historical bas reliefs, etc. The Canopy, surmounting, is elaborately painted in fresco, by Brumidi, and contains an immense number of allegorical and historical figures. Naturally the next objects of interest will be the

Senate and Representative Chambers, the former situated in the north wing (or "extension"—new part of the building) and the latter in the south wing. They are both large, with good accommodation for spectators (in the Strangers' Galleries) elaborately finished, lighted from above by hidden gas-burners through ground glass, and extremely well ventilated, but with their impressiveness materially marred by the low, flat ceilings. Both are reached by elaborate and costly stairways, really among the most notable features in the building; and it may be said of both that, with whatever faults of construction, they are among the best of their class, in the world. Next in interest is to be visited the

Supreme Court Room, a large semi-circular apartment in the north wing, with busts of former Chief-Justices Jay, Rutledge, Ellsworth and Marshall; and beneath it the Old Supreme Court Room, now the *Law Library*, with a fine collection of books in the higher branches of jurisprudence, and some peculiarities in the architecture of the room, commanding surprised attention. The

Library of Congress, with some 90,000 to 100,000 volumes (now accumulating very rapidly, as copies of all works published in America must be deposited there, to secure copy-right—as in the British Museum), and an immense number of valuable documents and manuscripts,—is in the western portion of the main building, and shows fire-proof bookcases and all appliances to guard against the recurrence of fire, which has twice destroyed previous collections (1814—war; and 1851, accidental.) But perhaps quite as interesting as any of the apartments named, is the

Old Hall of Representatives, in the south wing of the centre building, semi-circular, with panelled ceiling and cupola, row of splendid columns in Virginia green-stone; and containing, among other objects of interest and value, Vanderlyn's "Washington," a full-length of Lafayette, Franzoni's statue of History, statues and busts of Washington, Kosciusko, Lincoln, Johnson, &c. Here, also, have spoken nearly all the great legislators of America in the past, making the place historically memorable. Opening from this into the corridor, may be seen the

Bronze Columbus Door, modelled by Rogers and cast at Munich (where the model remains), representing various scenes in the life and death of the discoverer, and considered among the best of contemporary works.

Many Other Apartments of interest may be visited in the Capitol, among them the President's and Vice-President's Room, the Speaker's, Senators', Reception, &c., and some of the Committee Rooms, in the latter of which will be found displayed quite as much luxury (not always in the best taste) as can be found in any other portion of the structure.

[Admission to the Capitol, and access to most of its rooms, every week-day, 10 to 3. Congress generally in session, from early December to 4th March, in the years with odd numbers: in those with even numbers, (as 1872) the sessions often continuing far into the summer and even later. Admission to the Congressional Sessions, without card, and only limited by the capacity of the large galleries. In connection it may be well to say that the same hours (10 to 3) and the same freedom from routine or special application, apply to most of the Department buildings at Washington.]

Next in importance to the traveler, of the buildings of Washington, is the

PRESIDENT'S HOUSE (familiarly known as the "White House," especially in political parlance), situated on the high ground at the opposite or north-western extremity of (main) *Pennsylvania Avenue* (principal drive and fashionable promenade of the city). It is of white stone, as its name indicates, has a colonnaded front but little architectural merit, stands near the Potomac and commands a fine view of that river and the opposite shore. It contains some handsome and well-appointed rooms, the *East Room* being the most notable; but the location is not considered healthy in summer, and the Presidential family does not often steadily occupy it throughout that season. [Calls, without ceremonial or previous introduction, are generally received by the President every week-day, 10 to 1, except those devoted to Cabinet meetings or other special appointments. No court-dress necessary or proper. Levees, during the Congressional Season, fortnightly; and weekly receptions, generally on Saturday mornings, by the Lady of the White House, with the President present.]

THE PATENT OFFICE, after those named, is undoubtedly the most interesting place of visit in Washington, for its massive architecture and on account of its extraordinary collection of mechanical and labor-saving implements, in which it has no equal in any country. It is located on F street (many of the Washington streets being thus designated by letters), between Seventh and Ninth streets. The *Model Room*, occupying one entire floor, is divided into four halls, of which the East Hall is occupied by practical models; the West Hall by rejected ones; the South Hall (with handsome frescoed ceilings) by personal effects of Washington, other Revolutionary relics, (sword of Washington and cane of Franklin, among others) medals and treaties with, and presents from foreign powers, Powers' Statue of Washington, &c., forming a unique, most valuable and interesting collection. Near the Patent Office is to be visited the

General Post Office, an imposing Corinthian structure, with the internal arrangements commanding a certain degree of attention, and some valuable records of Franklin. Also, the City Post Office, in same building. The

State, War and Navy Departments have buildings near each other and near the President's House, on Pennsylvania Avenue. Little of interest is to be found in either, except the Library of the State Department, and the Collection of Relics of the War and Navy. Of much more importance to the visitor is the

Treasury Department Building, on Fifth street, immense in size and of some architectural merit;

while the details of *Paper Money Printing* [admission by order from the Secretary of the Treasury—easily obtained], carried on in the upper and lower portions of the structure, are worthy of close attention from their extent and completeness. The

Smithsonian Institute (founded by the late James Smithson, Esq., of England), stands in extensive and highly ornamented grounds, called the Mall, west of the Capitol, and south-east of the President's House. It is of large-extent, built of red sandstone, Norman in architecture, and has nine towers, of irregular heights. It contains an immense library-room, picture-gallery, lecture-room, laboratory, etc., and is already doing a noble work in the advancement of science. At no considerable distance from this, stands the

Washington Monument, intended to be one of the largest in the world, but thus far simply one of the largest failures, the funds to complete it from its present height of 170 feet to the contemplated 600, not being forthcoming. It is at present no monument but a curiosity. The

National Observatory stands on Western Pennsylvania Avenue, half way between the President's House and Georgetown. It has a large transit and some other fine instruments, astronomical library, clock, etc.

The Public Grounds of Washington are principally comprised in the *Mall*, on the banks of the Potomac, at and near the Smithsonian Institute;

the *Capitol Grounds*, before spoken of; and *Lafayette Square*, near the President's House (with Clark Mills' equestrian statue of Jackson). Principal Churches: the *Epiphany*, (Epis.) G. Street, near Thirteenth; *Trinity* (Epis.) Third street; *St. John's* (Epis.) Lafayette Square; *Presbyterian*, Four-and-a-half Street; *St. Aloysius* (Cath.) near the Capitol; *Foundry* church, (Meth. Epis.) Fourteenth street. Principal Theatres: the *New National* and *Wall's Opera House*. (*Ford's Theatre*, once a favorite, and the scene of the assassination of President Lincoln, April, 1865, is to be seen as a curiosity, but has never since been opened as a place of amusement). Public Hall: *Lincoln Hall*. Prominent Hotels: the *Arlington; St. Cloud; Howard; St. James'; Willard's*.

Suburban and other excursions from Washington, will include those to the *Soldiers' Home* (Military Asylum), three miles north of the city, and favorite resort of Presidential families in summer; the *Arsenal*, Greenleaf's Point, junction of Eastern Branch and Potomac, with interesting collection of ordnance. To the

Navy Yard, with ship-houses, an armory, etc., on the Eastern Branch, short walk south-east of the Capitol. To the

Congressional Cemetery, lying on the Eastern Branch, north-east of the Capitol, with many scores of monuments to Members of Congress who have died at Washington, and specially notable monuments of interest, to William Wirt, George Clinton, Elbridge Gerry, and others; also *Glenwood*, rival cemetery of much beauty, lying north of the Capitol. To the

Long Bridge, crossing the Potomac to the Virginia shore, from near the Mall, to Alexander's Island, and computed to have carried over half a million of troops during the war of the secession. To

GEORGETOWN, a handsome suburb, lying at the West, beyond Rock Creek, with fine view from the *Heights* and much interest in the *Aqueduct*, carrying the Chesapeake and Ohio Canal over the Potomac; also, *Georgetown College* (Jesuit), at the west end of the town, with library, museum, observatory, etc.; the *Convent of the Visitation*, Fayette street; Asylum for Children; *Oak Hill Cemetery* (with handsome Chapel, fine monument to M. Bodisco, etc.) To

Arlington House, former mansion of George Washington Parke Custis, stepson of Washington, and later of General Robert E. Lee, of the Confederate service, but the property since occupied as a Freedman's Village, and most of the rare Washington and other relics carried away, though the place must always retain a certain historical interest. To

Little and *Great Falls*, on the Potomac, above Georgetown, with handsome scenery and specially fine fishing. [All the last named places are most conveniently reached by carriage]. To

ALEXANDRIA, old town of Virginia, on the Western side of the Potomac, seven miles below Washington—once of heavy commercial importance, but now decayed. It has interesting historical reminiscences, in the fact that Gen. Braddock's unfortunate expedition to Fort Duquesne, which brought Col. Washington to notice, was fitted out here; also in his pew in Christ Church, still preserved, and other relics of the Father of his Country. It has a later and melancholy interest as the spot (at the Marshall House) where Col. Ellsworth, the Zouave, and his slayer, Jackson, were both shot, in May, 1861. It has also a Museum, Court House, Theological Seminary, etc. [Reached from Washington by special boat here and to Mount Vernon; or by regular daily boat on way to Aquia Creek, Fortress Monroe, and Richmond. Also, by rail or road. Railway connection, north to WASHINGTON; south to *Aquia Creek*, RICHMOND, etc.; northwest to *Leesburg*, HARPER'S FERRY, *Chambersburg*, etc.; west and southwest (through a succession of the early battle-fields of the secession war), to *Fairfax Court House, Manassas Junction* (battle-field of *Bull Run* in immediate neighborhood) connection northwestward to *Strasburg, Winchester*, etc.), *Warrenton Junction* (for Warrenton), *Culpepper, Orange Court House, Gordonsville*, CHARLOTTESVILLE (seat of the *University of Virginia*, and with "Monticello," home of Thomas Jefferson, lying near), *Summit*, and other stations, to the

WHITE SULPHUR SPRINGS, first of Virginia watering places, and one of the most popular of general resorts—lying on Howard's Creek, near the Western base of the Alleghany Mountains, which range is crossed in the latter portion of the transit. Virginia has many sulphur and other mineral springs; but the traveler without full leisure may take the waters of the White Sulphur, their well-kept hotels, and their beautiful natural surroundings, as affording the best type of the watering-places of the South]. To

MOUNT VERNON, old residence and burial-place of Washington, lying on the west bank of the Potomac, eight miles below Alexandria. It is, to Americans, unquestionably the most sacred of places on the continent, and only less interesting to those from other lands. Though somewhat decayed, the *House* (now the property of the Nation, through the labors of Edward Everett and the ladies of the "Ladies Mount Vernon Association"), commands a beautiful view of the river, is in fair preservation, and contains many valuable relics, among others, pictures and furniture of Washington, the Key of the Bastille, presented to him by Lafayette, etc. *The Tomb*, of brick, stands near the house, under heavy shade, with an open grated doorway through which the sarcophagi of Washington and his wife are seen; with other tombs of the family visible without; the archway of the tomb bearing the simple inscription: "*Within this enclosure rest the remains of General George Washington.*" Not far distant is the original Tomb, now crumbling to dust. [Reached from Washington by boat; or by boat to Alexandria, and drive, or by drive. If by boat, with view of and stop at old *Fort Washington*, once a fortification of some consequence, on the eastern side of the Potomac, between Alexandria and Mount Vernon.

Other Excursions from Washington, those to *Bull Run*, scene of the first important battle of the secession [reached from *Alexandria*, by Manassas Junction], and other fields of the late conflict. Also, nearly the same, in different directions, with obvious variation of railway routes, as those from Baltimore—see close of Division B, this route.

Division E.

WASHINGTON, TO AND AT RICHMOND, VA.

Leave Washington by morning boat on the Potomac river, to *Alexandria*, Va., on the right, [See previous Division]; thence by *Fort Washington*, on the left [See same], and *Mount Vernon*, on the right [See same]; to

Aquia Creek, small village deriving its only importance from this transit, at the junction of the Creek of that name with the Potomac. Thence rail, on the Richmond, Fredericsburg and Potomac road, to

FREDERICSBURG, on the right bank of Rappahannock River—old town of importance in early Virginia history, and especially notable from the fact that GEORGE WASHINGTON was born in the immediate neighborhood. This event, so important to the Western World and indeed to all mankind, took place upon what has long been known as the Wakefield Estate, at an inconsiderable distance from the town, within the limits of Westmoreland county; and though the birth-place has long been destroyed, the spot is commemorated by a stone slab erected there by George W. P. Custis, step-son of Washington, in 1815, and bearing the brief inscription: "*Here, the* 11*th of February,* (*O. S.*) 1732, *George Washington was Born.*" The mother of Washington resided, late in life, at Fredericsburg, and died and was buried there; her monument, in the outskirts of the town, inaugurated by President Jackson in 1833, still shamefully remaining unfinished. The house is still pointed out, at the corner of Lewis and Charles streets, where she saw her distinguished son for the last time. Fredericsburg has also a later celebrity, as the scene, and in the neighborhood, of a considerable amount of the fighting of the secession war, in 1862, '63, and later; and the country in the vicinity has by no means recovered from the devastation of those conflicts. Fredericsburg, by *Milford, Chester, Sexton's Junction* [connection westward, by Chesapeake and Ohio Railroad, to *Gordonsville, Staunton,* and *White Sulphur Springs,*] to

RICHMOND, on the James River, capital of the

State of Virginia, and ever memorable as the later seat of the Confederate Government, and the object of an investiture and siege by the United States forces, that seemed literally hopeless and interminable. It lies on the left or north bank of the James, at the Lower Falls, or end of that series of rapids extending six miles above and supplying the city with the needed water-power for its many flour-mills, tobacco and other manufactories. The most conspicuous object in the city, from the height of its position as well as from other causes, is

The Capitol, located on Shockoe Hill, a considerable elevation, and thus looking down upon the major portion of the city. It is Greek in the character of its architecture, with porticos, and a tall, narrow dome, and is generally impressive in effect, though the details are by no means faultless. It stands in a public square elevated as already named, and commands a fine view, especially from the portico or dome, over the James River, its islands, and a wide stretch of country. Internally, there is not much of interest in the legislative halls; the principal attraction centering in the splendid marble statue of Washington, by Houdon, considered the best extant, standing in the central hall, under the dome—and in the historical reminiscences, now so varied, inevitably clustering round the principal place of direction of the short-lived Confederacy.

Other Principal Buildings, worth visit: *Richmond* and *St. Vincent Colleges;* the *City Hall, Custom*

House and *Penitentiary;* and, as special objects of interest connected with the war, *Castle Thunder* and the *Libby Prison*. Also may be visited with profit, some of the many *Flouring Mills*, in which some of the best wheat in the world is prepared. Leading Churches: *St. John's* and the *Monumental*, with many others only less interesting. Other objects of interest: the old *Lines of Fortification* defending the city during the siege; the *Rapids* (or Falls of the James); the entrance of the James River and Kanawha Canal, etc. Leading Hotel: the *Ballard*.

[Principal railway connections from Richmond: northward, by routes just traversed, to WASHINGTON, etc.; eastward to the *White House* and Chesapeake Bay; southward, by Petersburg and Weldon road, to *Weldon* and *Wilmington* (N. C.); south-westward, by Richmond and Danville road, to *Greensboro*, (N. C.), and *Columbia* and CHARLESTON (S. C.); also south-westward, by South Side and Tennessee roads, across the Alleghanies to *Knoxville*, (Tenn.) and other places in extreme south and west. (See routes immediately following.)]

ROUTE NO. 10.—SOUTH-WESTERN (SEMI-SKELETON.)

RICHMOND, BY RALEIGH, WILMINGTON, COLUMBIA, CHARLESTON, ATLANTA, MONTGOMERY AND MOBILE, TO NEW ORLEANS.

Richmond by rail on Petersburg and Weldon road, to PETERSBURG (with fortifications remaining, and many other traces of the struggle which had some of its fiercest and most destructive conflicts at and around it); thence by *Hickford Junction*, where Raleigh and Gaston road is taken, and by *Ridgeway Junction*; to

RALEIGH, capital of the State of North Carolina, on the Neuse River, and named after Sir Walter Raleigh. It has an imposing *State House*, handsome *Union Square, State Lunatic Asylum* and many other objects of interest. From Raleigh; by the North Carolina and Wilmington roads, to

WILMINGTON, on the Cape Fear river, largest and chief commercial city of the State, with steamers to New York; extensive exports of naval stores; some good public buildings; *Forts Fisher* and *Caswell* (bombarded during the war), etc. Wilmington, by Columbia and Augusta road to *Florence;* thence by North-eastern road to

CHARLESTON, principal city of South Carolina,

and one of the leading sea-ports of the South [may be reached by steamer direct from New York], as well as especially celebrated as having been the spot at which the first fighting of the secession occurred, and for a long time the stronghold of the Confederates and object of Federal siege. It lies at the confluence of the Ashley and Cooper rivers, has a fine harbor, and very strong fortifications, in *Forts Moultrie, Castle Pinckney*, etc., and also the ruins of the celebrated *Fort Sumter*. It has many good public buildings, though many were destroyed during the war, from which the city is only slowly recovering. Among the most interesting buildings are the *Old State House*, (now Court House), *New Custom House, City Hall, Orphan Asylum, St. Michael's Church* (with fine old tower), *Charleston College*, etc. Principal Public Ground: the *Battery*, at the harbor-side. Principal Cemetery: *Magnolia Cemetery*, considered the finest in the South. Leading Hotels: the *Mills House, Charleston* and *Pavilion*. [Near connection south-westward, by Charleston and Savannah road, to

SAVANNAH, principal town of the State of Georgia, on south bank of the Savannah river, with remarkably wide streets, fine shade, many notable public buildings, revolutionary and secession reminiscences, and considered one of the healthiest of the Southern cities. Principal Hotels: the *Marshall, Pulaski*, and *Scriven*. Connection from Savannah south-westward to *Tallahassee* and other towns of Florida.]

From Charleston, by South Carolina road, by *Branchville* and *Kingsville*, to

COLUMBIA, capital of South Carolina, beautifully situated on the Congaree river, with what is considered the handsomest *State-Capitol* in the Union, the *South Carolina College*, and many other attractions, though burned during the war, and only partially recovered. Leading Hotel: *Nickerson's*. From Columbia, by Columbia and Augusta road, to

AUGUSTA, capital of Georgia, and second town in the State; on the Savannah river, with *Powder and Cotton Factories*, a large *U. S. Arsenal* in the neighborhood, handsome *City Hall*, and many attractions as a residence. Leading Hotels: the *Augusta* and *Planters'*. From Augusta, westward, by the Georgia road, to

Atlanta, important railway town of Georgia, being at the intersection of the Georgia road west, the Atlantic road southward from *Chattanooga* and NASHVILLE, the Macon road south to *Macon*, etc.; and with a certain other interest in its siege during the war, and as the point of departure of Sherman, on his "March to the Sea." Hotel: the *National*. From Atlanta, by Atlanta and West Point and Montgomery and West Point roads, by *West Point*, to

MONTGOMERY, capital of Alabama, and for a time the seat of the Confederate Government, before removal to Richmond. It lies on the Alabama river, has a commanding site, a *Capitol* worthy of attention, and many other good buildings, though having several times suffered severely by fire. Prominent Hotels: the *Central* and *Exchange*. From Montgomery south-westward, by Mobile and Montgomery road, by *Pollard* (Junction: railway connection to *Pensacola*, handsome town on Pensacola Bay, near the Gulf of Mexico, with fine harbor, U. S. Naval Station, etc.—leading Hotels: *Bedell, Winter* and *St. Mary's Hall*); to

MOBILE, on the Bay of the same name, branch of the Gulf of Mexico. It is the most important seaport of Alabama, and, in spite of bad navigation, the second of the Great Cotton-ports of the Gulf. It has few public buildings of interest, but fine waterviews, extensive fortifications, and a romantic historic interest as the scene of Farragut's fearful " passage of the Forts" and lashing himself in the shrouds of his vessel in the midst of their fire. [Communication by steamers and sailing-vessels, to NEW ORLEANS, *Galveston*, and many other ports on the Gulf.] Leading Hotel: the *Battle House*. From Mobile, by Mobile and Texas road, to

NEW ORLEANS, largest city of the State of Louisiana, and first cotton port of the South, as well as entrepot for products coming down the Mississippi River, of which it lies at near the debouchure into the Gulf of Mexico. New Orleans, familiarly called the "Crescent City," from its shape on the river, used also to be called the "Paris of America," and has not quite lost all the characteristics of gaiety bestowing the name. It lies on land lower than the river, rendering necessary a great embankment, called the *Levee*, which also supplies both wharves and promenades, along which may be seen the most marked features of the city. Among the later notable events connected, were another "passing of the Forts," below, (*Forts Jackson and St. Philip*) by Admiral Farragut, and the occupation of the city by the somewhat-unpopular commandant, Gen. B. F. Butler. Among the most important buildings are the *Custom House*, Canal street, one of the largest in America; the *U. S. Branch Mint;* the *City Hall; Odd Fellows Hall; Masonic Hall; Merchants' Exchange; U. S. Marine Hospital*, etc. It has many fine churches, with the Roman Catholic *Cathedral of St. Louis* the most prominent; and of its public grounds the most notable are the *New City Park, Lafayette Square, Jackson Square,* etc. Most attractive Cemeteries: *Cypress Grove, Greenwood,* and *Monument* (soldiers'). There are two Monuments of interest: the *Clay*, on Canal street, and the *Jackson* (unfinished) on the Battle-field, below the city. One of the most interesting features of New Orleans is to be found in the *Markets,* which should be visited early in the morning, not only to observe the immense variety of articles on sale, but the negro, half-Spanish and half-French characters of dealers and customers. Principal Theatres: the *Opera House, St. Charles, Varieties* and *Academy of Music.* Prominent Hotels: the *St. Charles, St. Louis, St. James,* and *City.* Excursions may be made to the *Battle-*

Field, scene of Gen. Jackson's victory over Sir Edmund Pakenham, Jan. 8th, 1815, four or five miles below the city; to the *U. S. Barracks*, a little above; to *Lake Ponchartrain*, above the city (famous for fishing and shooting, in the season); to the *Delta* and the *Mouths of the Mississippi*, some twenty-five miles below.

[New Orleans has regular communication, by steamer, to NEW YORK; to *Havana* (Cuba); to *Galveston* (Texas); and nearly all important Gulf ports. Also by steamer up the Mississippi, to *Memphis, Cairo, St. Louis*, and all important towns on that river. Also by rail, by *Jackson* (Miss.) to *Memphis;* and thence to all towns and cities in the North, North-east or North-west.]

ROUTE NO. 11.—SOUTH-WESTERN (SKELETON.)
WASHINGTON OR RICHMOND, BY LYNCHBURG, KNOXVILLE AND CHATTANOOGA, TO MOBILE AND NEW ORLEANS.

Washington by rail, by *Alexandria;* and by Orange, Alexandria and Manassas road (by *Manassas Junction*), to *Charlottesville* (Junction—connection westward to *Staunton,* etc.); thence direct to Lynchburg. Or, Richmond by South Side road to *Burkville* (Junction—intersection with Richmond and Danville road, southward); thence direct by *Appomattox,* and other stations, to

LYNCHBURG, on the James River, and the James River and Kanawha Canal—important tobacco-depot and flourishing town. [Most convenient railway point, from which to reach, by canal-packet or carriage, those great natural curiosities, the *Natural Bridge* and the (Mountain) *Peaks of Otter.*] From Lynchburg, by Virginia and Tennessee Road, by *Bonsack's* [stage connection to *White Sulphur* and other Springs]; *Big Tunnel* [passage of the Alleghany Mountains; horse-car connection to *Alleghany Springs*]; *Bristol,* and other Stations, to

Knoxville, important town of the State of Tennessee, on the Holston River, with the *University of*

East Tennessee, many railway connections, and much popularity as a place of residence. Knoxville to

CHATTANOOGA, on the Tennessee river, near the boundaries of Alabama and Georgia, and one of the most important railway centres of the south-west; but additionally celebrated, since the war, for the battles of *Chickamauga* and *Lookout Mountain*, fought in the immediate neighborhood. In the vicinity of the Lookout (easily visited from Chattanooga,) is to be found scenery of equal grandeur and loveliness. Hotel: the *Crutchfield House*. Chattanooga, by the Alabama and Chattanooga road, by *Tuscaloosa* and other important stations, to

Meridian, railway town of the State of Mississippi, [with connections east to *Montgomery*, west to *Jackson* (capital of the State), north of *Memphis*, etc.] From Meridian, by the Mobile and Ohio road, direct to MOBILE and New Orleans, as in Route No. 10.

ROUTE NO. 12.--WESTERN.

NEW YORK TO PHILADELPHIA, HARRISBURG, PITTSBURG, WHEELING, COLUMBUS AND CINCINNATI, BY PENNSYLVANIA CENTRAL ROAD AND CONNECTIONS.

Division A.

NEW YORK TO PHILADELPHIA, OR MANTUA JUNCTION.

Leave New York (as by Route No. 8) by the New Jersey road, by *Jersey City, Newark, Elizabeth, Rahway, New Brunswick, Trenton*, etc., to PHILADELPHIA, if for stop at that city; if for through passage to the West, without stop at Philadelphia, New York by the same towns to MANTUA JUNCTION, where close through-connection is made.

Division B.

PHILADELPHIA, OR MANTUA JUNCTION, TO AND AT HARRISBURG.

Leave Philadelphia (West Philadelphia), or Mantua Junction, if without stop at Philadelphia, by rail, by the Pennsylvania Central road; by *Downington* [connection northward for *Waynesburg*]; by *Coatesville* [connection northward for *Reading*, southward for *Wilmington*], etc.; to

LANCASTER, pleasantly situated on the Conestoga Creek, in a fine agricultural section; seat of *Franklin* and *Marshall College*; with *Court House* and other creditable buildings, and interesting series of *Canal Locks* in the neighborhood, at mouth of the creek. Was for some years, at beginning of the century, the seat of government of Pennsylvania. Leading Hotels: the *City*, and *Michael's*. Lancaster, by *Branch Intersection* [connection northward to *Reading*, southward for *Columbia, York*, etc.], *Mt. Joy* and *Middletown*, to

HARRISBURG, capital of the State of Pennsylvania, on the east bank of the Susquehanna river (originally "Harris' Ferry" over that river). The most notable building is the *Capitol*, on high ground, with fine view from the dome, with State Library, Legislative Chambers, etc. Also should be visited, the *Court House;* the *Old Harris Mansion;* and some of the extensive Iron and Steel Works in the vicinity; as well as the remains of the earthworks thrown up to defend the city against the Confederates, with burning of bridges, in 1863. Principal street; *Front Street*. Principal Public Ground; *Harris Park*. Prominent Hotels; the *Lochiel, Jones House*, and *Bolton's*. [Important railway connections from Harrisburg: by Lebanon Valley road, east to *Lebanon* and *Reading;* by Northern Central road, southeast to BALTIMORE, etc.; by the same road northward to *Elmira* and the Erie Railway and its connections; by Cumberland Valley road, southwestward for *Carlisle, Chambersburg*, etc.; by Philadelphia and Erie road, northwestward for *Williamsport, Corry, Erie* and Oil Regions.]

Division C.

HARRISBURG TO AND AT PITTSBURG, WHEELING, ETC.

From Harrisburg, continuing by Pennsylvania Central road; by *Lewistown* [connection northward for *Milroy*, northeastward for *Sunbury*, etc.]; by *Tyrone* [connection northeastward to *Lock Haven* and the Philadelphia and Erie road, northwest to *Clearfield*, etc.]; to

ALTOONA, at the commencement of the ascent of the Alleghany Mountains; great locomotive-shop of the Pennsylvania Central Company; and surrounded by magnificent mountain-scenery, making a sojourn very pleasant in summer. Hotel: the *Logan House.* [Spur southward to *Martinsburg*, and stage thence to *Bedford Springs.*] From Altoona should be made, by daylight, to enjoy the fine scenery, the

Railway ascent of the Alleghanies, with features quite as grand as most of the Alpine rail-routes, and double power necessary in drawing up the trains. An immense *Tunnel*, nearly three-quarters of a mile in length, is passed through before reaching the summit, at

CRESSON (Cresson Springs), a popular summer-resort, on account of its elevation and healthful air. Hotel: the *Mountain House.* [Spur northward to *Ebensburg.*] From Cresson the *descent of the Alleghanies* is made, *without the use of steam*, the speed being regulated by brake-power; to *Conemaugh Station*; and to *Johnstown*, with the extensive Cambria Iron Works in the neighborhood, and heavy manufactures. Hotel: the *Scott House*. From Johnstown, by *Blairsville* [connection northward for *Indiana*, northwestward to *Freeport* and points on Allegheny Valley road]; by *Greensburg*, etc., to

PITTSBURG, on the head-waters of the Ohio river, at the confluence of the Alleghany and Monongahela, and on the spot once occupied by the old Fort Duquesne, scene of the defeat of General Braddock ("Braddock's Field") in the English and French colonial war. In Pittsburg (named after William Pitt); in *Alleghany City* (across the river, with no less than 5 connecting bridges); in *Birmingham, Lawrenceville* and other suburbs—is concentrated the most extensive chain of manufactures, in iron, glass, steel, brass, wooden-wares, etc., on the American Continent, and scarcely if at all second, in those regards, to any city of the world. It is also an immense coal and oil centre, with the most extensive refineries of the latter; and the other industries are almost innumerable; while the coal-smoke of so many factories gives to Pittsburg the unenviable distinction of having the worst-clouded and dirtiest atmosphere in America.

After the *Manufactories*, the most important objects to the visitor, in Pittsburg, are the *Roman Catholic Cathedral; Presbyterian* and *Baptist Churches; Court House; Custom House;* (with Post Office); new *City Hall; Mercantile Library Hall*, etc.; and in Alleghany City, the *Theological Seminaries*, Western Penitentiary, etc., and more elegant residences of the citizens. There are no less than four Cemeteries: the *Alleghany; St. Mary's; Hilldale;* and *Mt. Union.* Leading Hotels: the *Monongahela, Union, St. Charles,* and *Merchants'.*

[Railway connections from Pittsburg are very general. Northward, by the Alleghany Valley road, to *Venango, Oil City,* and the Oil Regions generally; eastward by route just traversed; westward, to places named, by the Pittsburg, Cincinnati and St. Louis and Pittsburg, Fort Wayne and Chicago roads; northwestward to *Cleveland,* by Cleveland and Pittsburg road, and to *Erie* by the Erie and Pittsburg road; southeastward to *Cumberland, Harper's Ferry,* etc., by the Pittsburg and Baltimore and Washington road. There is also steamboat communication, down the Ohio river to *Wheeling,* and thence to *Cincinnati,* during the open season.]

From Pittsburg, by Cleveland and Pittsburg road to

WHEELING, West Virginia, a large and important town, lying at the debouchure of Wheeling Creek into the Ohio river—with manufactures of the same character as those of Pittsburg, only second to them in extent. Apart from its *Manufactures,* the two greatest points of interest are the *Wire Suspension Bridge* of the National Road, with 1.000 feet of span; and the new and splendid *Railway Bridge.* Oil and coal trade also immense, as at Pittsburg. [Railway connection southeastward by the Baltimore and Ohio road to *Harper's Ferry, Baltimore* and *Washington;*

eastward to *Pittsburg* and northwestward to *Cleveland* and Lake Erie, by the Cleveland and Pittsburg road. Or, by Baltimore and Ohio road, from Wheeling by *Belle Air ;* and ZANESVILLE, thriving and handsome town on the Muskingum River, with immense water power and fine railroad-bridge —[connection southwestward, by Cincinnati and Muskingum Valley road, to *Cincinnati*]: to *Newark, Columbus,* etc. Steamboat communication to *Pittsburg,* and down the Ohio to CINCINNATI during the open season. Hempfield railway will connect directly to Pittsburg when completed.]

Division D.

PITTSBURG TO AND AT COLUMBUS, OHIO.

From Pittsburg, by rail, on the Pittsburg, Cincinnati and St. Louis road (or from Wheeling by rail to same point); by

STEUBENVILLE, pleasant village on the Ohio river, county seat of Jefferson County, with many manufactures and fine scenery in the neighborhood; by *Mingo Junction* [connection northwestward to *Cleveland,* eastward to *Rochester,* etc.]; by *Dennison,* COSHOCTON, and *Dresden Junction* [connection south to ZANESVILLE, etc.]; to

NEWARK, handsome and thriving town on the Licking river, with extensive railway connections: roads to *Sandusky* and Lake Erie, to *Zanesville* and the south, intersecting. From Newark to

COLUMBUS, on the Scioto River, capital of the

State of Ohio, and one of the most important towns of the state. It is beautifully laid out and very handsomely shaded; *Broadway*, its main street, being considered unsurpassed in any land. The *Capitol* is nearly new and very imposing (lying on the elegant public ground, *Capitol Square*); and there are, of other public buildings of interest, the *City Hall; U. S. Arsenal*, with fine high grounds; *State Penitentiary; Central Ohio Lunatic Asylum* (building, in place of that burned in 1868); *Blind* and *Deaf and Dumb Asylums; Starling Medical College; St. Mary's Female Seminary*, etc. Also worthy of attention are the *Holly Water Works*, with steam raising-power. Other Public Grounds than the one already named and the fine ones surrounding most of the public buildings: the *City* and *Goodale Parks*, and those of the *Franklin Agricultural Society*. Most popular Cemetery: *Green Lawn*. Theatre: the *Opera House*. Hotel: the *Neil House*. [Railway conections extensive: eastward by the route just traversed; northward by the Cleveland, Columbus, Cincinnati and Indianapolis road, to *Toledo, Cleveland*, etc.; south-westward to *Cincinnati* (as see route following); south-eastward to *Athens* and the Baltimore and Ohio road; etc.]

Division E.

COLUMBUS TO AND AT CINCINNATI.

From Columbus, by the Little Miami road; by *London*; by *Xenia*, very handsome town, with water-power and manufactures, on the Little Miami river [connection westward for DAYTON and *Richmond*]; by *Morrow* [connection east with the Cincinnati and Muskingum Valley road]; by *Loveland,* [connection east by the Marietta and Cincinnati road, for Marietta, and the Baltimore and Ohio road]; to

CINCINNATI, on the Ohio river, called the "Queen City," principally built upon two terraces sloping back from the river; while opposite it, and divided from it by the Ohio river, are the large towns of *Newport* and *Covington,* in the State of Kentucky. Cincinnati is considered very handsome, and, though hot in summer, healthy; and it ranks well in manufactures, and commercially among the first of western cities.

Among the public buildings of prominence are the *Custom House* (with Post Office attached), on Fourth Street; the *City Hall* (with neat grounds), Plum Street; the *Court House,* Main Street; *Cincinnati College,* Walnut Street; *St. Xavier's College* (Catholic) Sycamore Street; *Convent of Notre Dame,* Sixth Street; *House of Refuge,* north of the city; *City Workhouse,* near the latter; *Cincinnati Hospital,* Twelfth Street; etc. Principal Churches: *St. John's* (Epis.); *St. Paul's* (Meth. Epis.); *First Baptist; St. Peter's Cathedral* (Catholic); *First Presbyterian,* etc., though with many others creditable. Places of Amusement: the *National* and *Wood's Theatres; Pike's Music Hall; Melodion; Gymnasium; Queen City Skating Rink,* etc.

Prominent Hotels: the *Burnet, Spencer, Gibson, St. James, Carlisle*, etc.

Public Grounds: *Eden Park*, east of the city, elevated and with *fine view*; *Fountain Square*, with magnificent bronze fountain lately presented by Mr. Henry Probasco; *City, Lincoln, Washington* and *Hopkins Parks*. Cemeteries: *Spring Grove*, one of the handsomest in the West, northwest of the city, with splendid avenues of approach, and a fine soldiers' monument; *St. Bernard, Wesleyan*, and others minor. Other Objects of Interest: the great *Suspension Bridge* over the Ohio, with longest span in the world; the *Licking Bridge*, also a suspension, and only less remarkable in length; the *Railroad Bridge* (new); remains of entrenchments thrown up during the Confederate "siege"; the *Levee*, along the river, with steamboat-landings and a very fine idea of the industry of the city; steamboat-building-yards; and many of the very extensive Manufactories, with diversified products.

[Railway connections: eastward, by route just traversed—also, by Marietta and Cincinnati, and Chesapeake and Ohio roads to *Richmond*, etc.; northeastward by the Little Miami and other roads, to *Cleveland, Sandusky*, etc.; northward, by Cincinnati, Hamilton and Dayton road, to *Toledo;* northwestward, by same road and connections, to CHICAGO; westward, by Ohio and Mississippi road, by *Vincennes* to ST. LOUIS and the Mississippi river.

Also, steamboat transit on the Ohio river, to all points on that stream, to *Cairo* and the Mississippi.]

ROUTE NO. 13.—WESTERN.

NEW YORK TO EASTON (PA.), HARRISBURG, PITTSBURG, FORT WAYNE AND CHICAGO; BY THE ALLENTOWN ROUTE.

Division A.

NEW YORK TO EASTON, HARRISBURG AND PITTSBURG.

Leave New York, by boats of New Jersey Central Railroad, from foot of Liberty Street, to *Communipaw* (lower Jersey City); thence by cars of that road, by *Bergen Point,* and over *Long Bridge* across Newark Bay; to *Elizabethport* (station—at the left the town and great coal depot of that name); and to ELIZABETH [intersection with New Jersey road, eastward to NEW YORK, and westward to PHILADELPHIA: See Route No. 8, New York to Philadelphia]. Elizabeth, by other stations, to

PLAINFIELD, pleasant village and favorite summer-residence, lying at near the foot of the Orange Mountains, a minor spur of the Blue Ridge, and with a remarkable eminence at a short distance to the north, called "Washington's Rock," from which that general is said sometimes to have watched the movements of the British forces. By *Bound-Brook,* on the Raritan River, and at the opening of the Valley of the same name, to

SOMERVILLE, county seat of Somerset County, very handsomely situated, with fine quiet scenery in the neighborhood, some copper and iron mines, and much general prosperity. [Connection by South Branch Road to *Flemington* and *Lambertville*.] By other stations to the

High Bridge, (or rather very long and high *embankment*) over the South Branch of the Raritan River, with fine view in crossing, and large Iron-Works in the neighborhood, for railroad-founding. Very soon is reached

Hampton Junction [connection with the Delaware, Lackawanna and Western Railway, for the *Delaware Water Gap, Scranton*, the Coal Regions of Pennsylvania, and the Erie road at Binghampton.] By *Bloomsbury* and other stations, to

PHILIPSBURG, on the New Jersey side of the Delaware, with heavy iron-manufactures, and three bridges connecting it with *Easton*, on the opposite side of the river. Also with important railway connections, for Central New Jersey, *Philadelphia*, and northward to the *Delaware Water Gap* and the Coal Regions. Crossing the river by bridge, the train reaches

EASTON, Pennsylvania, on the western bank of the Delaware, at the double junction of the Lehigh and the Bushkill, and one of the most important of the coal and railway centres of the two adjoining States, with extensive mills, distilleries and general manufactures. It is also the seat (on an eminence known as *Mount Lafayette*, at the east of the town) of *Lafayette College*, a flourishing and well-endowed institution, rapidly growing in influence. [Extensive railroad connection, apart from the line of route being traversed; as in addition to that mentioned by both the New Jersey Central and the Delaware and Lackawanna to NEW YORK, it has also connection northward to the *Water Gap* and the Coal Regions; southward to PHILADELPHIA; the Lehigh Valley road northward to *Pittston*; the Lehigh Canal in the same direction; and the Morris Canal through the State of New Jersey to Raritan Bay.]

Leave Easton by Lehigh Valley road; by BETHLEHEM, pleasant town on the Lehigh river; seat of *Lehigh University*; and long celebrated as the principal abode of the Moravians, or United Brethren, in the United States. [Connection southward with North Pennsylvania road for PHILADELPHIA, and northward for *Scranton* and the Coal Regions.]

From Bethlehem, by *E. Penn. Junction* [connection with East Pennsylvania road]; to

ALLENTOWN, handsome town on high ground near the Lehigh river, with large iron and other manufactures, and much charm as a residence. Has *Big Rock* and several popular mineral springs in the neighborhood. Hotel: the *American*. [Connections, northward by the Lehigh Valley road to the Coal Regions and the Erie road; westward by present route to *Reading*, etc.]

Allentown, by East Pennsylvania road, to

READING, large and important manufacturing town on the Schuylkill river, especially notable for mills, iron-furnaces and railroad work. It has a high and handsome sloping location, with a considerable eminence, *Penn's Mount*, near, commanding fine view. It has a noble *Court House*; two or three *Churches* of especial beauty; and *Mineral Springs* in the neighborhood, with hotels of popular resort. Leading Hotel (in the town), the *Mansion House*. [Connection, northward, to *Catawissa* and *Hazleton* (Coal Regions), and southeast to PHILADELPHIA by the Philadelphia and Reading road.] Reading to

LEBANON, on the Swatara Creek, county seat of Lebanon County, and a prosperous town, with the most immense (Cornwall) Iron Ore Beds in the neighborhood, known to exist in the world; also Copper Ore in large quantities, and Marble. From Lebanon to

HARRISBURG. (For notes on Harrisburg, see previous Route, No. 12.)

[Harrisburg to *Pittsburg*, as by Route No. 12, preceding; whether for *Chicago*, *Cincinnati*, or *St. Louis*.]

Division B.

PITTSBURG TO FORT WAYNE AND CHICAGO.

[At Pittsburg, previous route, No. 12, may be pursued, from that point to *Columbus* and *Cincinnati*, with extension to *St. Louis;* or other lines pursued (see that route) northward to the Lakes, southward to the Baltimore and Ohio road, etc.]

Leave Pittsburg, for CHICAGO or places on that line, by Pittsburg, Fort Wayne and Chicago road; by *Rochester* (small town on the Beaver Creek—with connection southwestward, by Cleveland and Pittsburg road, to *Wheeling* and west); by Homewood [connection northward for *Newcastle*, etc.]; by *Leetonia* [connection southward to New Lisbon, etc.]; by *Salem*, very handsome small town, with fine suburbs, and manufactures]; to

Alliance, important station [connection northward, by Cleveland and Pittsburg road, to *Cleveland;* southward to *Steubenville*, etc.] From Alliance, by *Canton* and *Massillon*, handsome manufacturing towns; by *Orrville* [connection northward to AKRON and *Cleveland*]; by *Mansfield*, another handsome manufacturing town [connections northward to *Sandusky*, southward to *Zanesville*, southwestward to *Dayton, Hamilton* and CINCINNATI]; to

Crestline, another important railway town. [Connections southward by the Cleveland, Columbus, Cincinnati and Indianapolis road, to *Columbus;* northward by the Sandusky road to *Sandusky;* northeastward to *Cleveland* by the Cleveland, Columbus, Cincinnati and Indianapolis road; northwestward to *Toledo* and *Detroit* by same road and connections]. From Crestline, by *Bucyrus*, thriving village on the Sandusky River; by *Forest* [connections north to *Cleveland*, south to *Cincinnati*]; by

Lima, manufacturing village on the Ottawa river [connections southward by Cincinnati, Hamilton and Dayton road, to *Dayton* and *Cincinnati;* northward by the same road to *Toledo* and *Detroit*] ; by minor stations, to

FORT WAYNE, Indiana, on the Maumee river—called the "Summit City"; because it lies at the highest point of the water-shed. It is passed through by the Wabash and Erie Canal, and is an important railway centre as well as a manufacturing town of promise. [Connections: northeastward to *Toledo,* Lake Erie and *Detroit,* by Toledo, Wabash and Western road ; westward by the same road and connections, to *Logansport* and thence to *Peoria* and towns of Central Illinois; northward to the Michigan Southern road, at *Waterloo,* etc.] From Fort Wayne, by *Columbia:* by *Warsaw;* by *Plymouth* [connection southward to INDIANAPOLIS, capital of the State, by Indianapolis, Peru and Chicago road]; by *Wanatah* [connection southward by the Louisville and New Albany road, to *Lafayette* and to LOUISVILLE, Ky.]; by Valparaiso, to that city which has furnished, both in fortune and misfortune, the best possible type of American capacities in either direction, and which has been, and will continue to be, quite as often in men's mouths as any other on the Western Continent—

CHICAGO, Illinois.

Division C.

CHICAGO AS IT WAS AND IS.

The history of the world furnishes no parallel to the change between the two words just given, the "was" and "is" of the lately-great and yet-to-be-greater city of Chicago. When a considerable portion of the material of this book was already in type, Chicago stood, as it had been for thirty or forty years growing up to be, the actual Queen City of the West, one of the most important of the Union, and the greatest grain mart and depot of the world, as well as one of the most important railroad centres of the entire continent. It lay on the western shore of Lake Michigan, at near the southern boundary of that Lake, at the entrance of the Chicago river into it. It had its first white settlement in 1804, by Col. John Kinzie; and so late as 1830 contained only 15 houses. It was incorporated as a city, seven years later, in 1837—the population at that time being 4,170. In 1843 this had increased to 7,580; in 1847 to 16,859; in 1850 to 28,269; in 1855 to 80,023; in 1860 to 109,263; in 1865 to 178,539; and in 1870 to the round figures of 300,000, while the suburban population was supposed to raise it to 350,000.

The site of the city was admirably chosen, on ground sloping up from the Lake, and with the

Chicago River, dividing into two branches, running through the entire city, at once adding to commercial convenience and healthfulness. Numerous costly bridges and many tunnels made the connection across the rivers. The city was divided into 20 wards, with nearly 800 streets. The river and the ship canals afforded many miles of excellent harbor, to which came vessels from all parts of the great lakes, with and for produce of every description. The grain warehouses were of such extent as to accommodate 8,000,000 to 10,000,000 bushels; while the yearly exportation had reached to about 60,000,000 to 65,000,000 bushels. It had also a most extensive trade in cattle and stock, the whole yearly number handled, reaching 2,500,000; besides provision, lumber, and transportation trade to immense amounts. Within the last year or two, on the completion of the Pacific Railway, Chicago had commenced to import her teas and silks direct from India, and was arranging a heavy trade in that direction. There were several of the largest and finest railway-depots on the Continent, to accommodate its immense connection as the very largest of the railway-centres. It had streets among the most elegant in the West, in Michigan, Wabash and other avenues: and many of the houses of residents were princely in their luxury. There were more than 200 Churches; 12 or 15 of the largest and finest of Hotels, some of them marvels of size and cost; an Opera-House and 5 Theatres of good class; an University; Medical Colleges; splendid public Parks and Cemeteries; and a wilderness of scientific, social and benevolent institutions, second to none in the Western World. In addition to this, and a feature unequalled elsewhere—a Tunnel had been constructed to a Tower two miles distant in the Lake, whence the water-supply of the city was derived.

Such, hastily sketched, was the position of Chicago on the 8th of October, 1871. On the night of that day, a fire broke out at near the centre of the city, though in an older and wooden portion. A fierce wind made vain all efforts for staying the flames, which extended on every side, and eventually acquired such force that buildings considered fire-proof could not resist the heat five minutes when attacked. By the morning of the 10th the city was literally destroyed—the whole central and business part of it entirely so. Some 10,000 buildings were burned; 500 to 1,000 persons are supposed to have lost their lives; 50,000 to 75,000 persons were rendered houseless; and the pecuniary loss is estimated to have reached $200,000,000. Chicago, the Queen of the West, was no more—*for the present.*

A most gratifying spectacle of general benevolence, however, has been shown, in connection with this great calamity, not only in all the cities of America, but in those of England and of all Europe; and the extremity of suffering has been materially relieved by contributions from all quarters, which will no doubt continue during the following months of helplessness and want. Meanwhile, the rebuilding of the city was commenced at once, and is already proceeding rapidly; and, though years must elapse before the terrible marks of the visitation are obliterated, all the industries of Chicago will soon be in full even if limited operation. Already, *all its railway facilities as a great centre, temporarily deranged, are again fully supplied;* and the thousands of visitors, who would have gone to see it as a curiosity of Western greatness and prosperity, will still do so, to see it in its prostration and rapid revival, with all the facilities of transit and accommodation that would have been originally enjoyed. It is impossible to say, at this early day, what Hotels will be in readiness to accommodate visitors, but certainly two or three of excellent class and capacity.

[Among the widely-extended railway connections of Chicago, are the following of most importance. Northward to *Milwaukie*, thence to *Green Bay*, and beyond, to the Lake Superior sections, by the Chicago and Northwestern road and its connections; northwestward to *Janesville*, MADISON (capital of Wisconsin), thence to *Prairie du Chien*, and to ST. PAUL and the *Falls of St. Anthony*, by the same road and its connections; westward to *Cedar Rapids* and *Des Moines* (Iowa), by the Iowa division of the same road; southwestward to *Burlington* (Iowa), and the Mississippi River, by the Chicago, Burlington and Quincy road: southward to ST. LOUIS, by the Chicago and Alton road; southward to *Cairo* (junction of the Ohio and Mississippi rivers) by the Illinois Central road; westward to *Omaha*, and thence to SALT LAKE CITY and SAN FRANCISCO, by the Chicago and Northwestern, Chicago and Rock Island, and other routes; southeastward to CINCINNATI, by the Columbus, Chicago and Indiana Central road; eastward to PHILADELPHIA and NEW YORK by the route just traversed—as also by *Toledo, Cleveland* and the Lake Shore road—as also (with all Canadian cities and connections) by *Detroit* and the Grand Trunk Railway of Canada. Also, steamers on the Lakes, to all important points, in the season.]

ROUTE NO. 14—WESTERN.

CINCINNATI TO LOUISVILLE (MAMMOTH CAVE), NASHVILLE, CAIRO, ST. LOUIS AND CHICAGO.

Division A.

CINCINNATI TO LOUISVILLE, NASHVILLE AND ST. LOUIS.

Leave Cincinnati by the Louisville, Cincinnati and Lexington road (from *Covington*—opposite side of the river); by *Walton, Sparta, Lexington Junction* [connection southward for LEXINGTON and *Ashland*, old home of Henry Clay, near it; and, by stage from *Eminence*, for Shelbyville]; by *Lagrange*, and *Anchorage* [connection with Shelby road], to Louisville. (Or, steamer down the Ohio from Cincinnati, in the pleasant season).

LOUISVILLE, located on the Ohio river, at the Falls and near the entrance of Bear-Grass Creek, is the largest and most important city of Kentucky. It is well located and shaded; and the views of the Falls, from various points of the city, are much admired. The most important trade of the city is in *tobacco*, of which it is one of the central marts: also extensively in flour, provisions, hemp, etc. The most interesting public buildings are the *City Hall*,

Court House, Custom House (with Post-office), *University Medical College, Masonic Temple, Blind Asylum,* the *Cathedral, St. Paul's Church,* etc. Principal theatre: the *Louisville.* Leading Hotels: the *Louisville,* and *National.* Principal Cemetery: *Cave Hill,* with many monuments of merit. At the opposite side of the river is JEFFERSONVILLE, Indiana [railway connection to Indianapolis].

[It is from Louisville that detour may be most conveniently made to visit the

Mammoth Cave of Kentucky, one of the most extensive subterranean passages in the world, and considered among the most interesting. Or, it may be taken on the way from Louisville to *Nashville,* as at present to be considered].

Leave Louisville by the Louisville and Nashville road, to *Cave City,* whence stage or carriage, 9 miles to the Cave (or, steamer from Louisville, on the Green River, to within 1 mile of the Cave—thence on foot). A Hotel, the *Cave House,* affords facilities for stoppage, during the exploration, which may be brief, but must, for any approach to thoroughness, require days of interest and toil, always accompanied by a guide, and with lights and means of relighting, without either of which it is not safe to enter. For particulars of the chambers, passages, and various parts of interest in this wonderful cave, which is believed to extend eight or nine miles back from the entrance, dependance may be made entirely upon the capable and instructed guides, procurable at the hotel or the entrance. Three other Caves—the *Indian*, *White's*, and *Diamond*, may be found in the vicinity: the two former with peculiarly handsome stalactite and stalagmite formations, miniatures of those in the great cave.

From Cave City (after return from the Cave) continue route by Louisville and Nashville road; by *Bowling Green; Memphis Junction* [connection southwest for *Memphis* and the Mississippi river]; by *Junction* [connection northwest by Henderson road to *Henderson*, and *Evansville* (Ind.); to

NASHVILLE, Capital of the State of Tennessee, and one of the most important cities of the middle southwest. It lies on the Cumberland river, on elevated ground, much of the city lying nearly or quite 100 feet above the water level, and being considered very healthy. The most prominent building, the *State Capitol*, is considered one of the finest in the Union, having admirable legislative halls, splendid material of native marbles, a tower, State Library, etc. Other buildings of prominence are the *University*, the *City Hall, Lunatic Asylum, Penitentiary*, etc. Theatres: the *Nashville*, and *Duffield's*. Prominent Hotels: the *St. Cloud*, and *Stacey*. Nashville has many handsome residences; and the tone of its society is considered equal if not superior to that of any other city of the South. Very near it may be visited the *Hermitage*, old seat of Andrew Jackson. [Extensive railway connections; east to *Knoxville*; southeast to *Stevenson*; southward, by

Nashville and Decatur and connecting roads, to *Montgomery* (Ala.), and thence to the Gulf Cities; northward, by Evansville road, to *Vincennes* and *Terre Haute;* southwestward to *Memphis,* by Nashville and North Western road, and connections; northwestward to *St. Louis,* etc.]

From Nashville, by North Western road, by *Waverley; Johnsonville; McKenzie* [connection southwestward, by Memphis and Louisville road, for *Memphis,* and for *Little Rock* (Arkansas)]; by *Paducah Junction* [connection north to *Paducah*]; to

Union City, where connection is made with the Mobile and Ohio road. By that road to

Columbus, on the Mississippi River; with *Belmont,* Missouri, opposite, connecting the route just traversed with the St. Louis and Iron Mountain road to *St. Louis.*

From Columbus the Mississippi may be ascended, by boat, to

CAIRO, modern town, very low-lying, on the point formed by the confluence of the Ohio and the Mississippi, with costly levee against inundations by the river, and much prominence as a steamboat port of the Mississippi, vessels from and to all ports stopping here to land and receive passengers and freight. Has some noble buildings; among the best, the *Custom House.* [Railway connection, from Cairo, or from *Mound City,* immediately above, with the Illinois Central Road, direct for CHICAGO. Or, steamboat may be taken for ST. LOUIS or any other point on the Mississippi. Or, steamboat may be taken for *Louisville* or CINCINNATI].

For St. Louis, from *Bird's Point* (opposite Cairo) to *Charleston ;* where the St. Louis and Iron Mountain road is taken. By *Glen Allen, Marquand,* and other stations, to *Bismarck* (where pause should be made, if time allows, to visit, by a spur of the same road, *Iron Mountain, Pilot Knob* and *Ironton,* with some of the most extraordinary developments of richness in iron mines, on the continent). Bismarck, by *Mineral Point* [spur to *Potosi*] and other stations, to ST. LOUIS.

Division B.

AT ST. LOUIS; AND BY SPRINGFIELD TO CHICAGO.

ST. LOUIS, Missouri, is one of the largest and most important cities of the West, focus of mercantile supply for a wide extent of country, virtual Queen of the Mississippi, and often spoken of as the point for removal of the National Capital. It lies on the west bank of the Mississippi river, at what is supposed to be about half-way between St. Paul, at the head of navigation on the Missouri, and New Orleans, at the mouth of the Mississippi. It occupies elevated ground, though uneven; has a very long extent on the river, and an imposing appearance from it. It has wide streets, with good shade; handsome parks; substantial residences; and one feature commanding unmixed admiration, in the *Levee,* at which the most immense number of steamboats can at any time be seen lying, loading, discharging, arriving and departing, observable at any one spot on the globe. *Front Street*, along the Levee, is one of the finest of mercantile and warehouse streets in the Union; while *Washington* and *Grand Avenues*, and *Fourth Street*, are among the most fashionable thoroughfares.

St Louis is singularly rich in Parks; the most notable being *Lafayette, Hyde, Laclede* and *Gravois Parks*, in the outskirts; *Washington* and *Missouri*, and smaller squares; with *St. Louis Park* just being commenced, to contain more than 3,000 acres and rival the *Fairmount* at Philadelphia. The *Fair Grounds* are also very beautiful and perfect, as well as popular, with an *Amphitheatre* for spectators, estimated to accommodate 80,000 to 90,000 persons; and the *Botanical Gardens* are considered the best on the continent. The Principal Cemeteries are the *Bellefontaine* and the *Calvary*.

Among the Buildings best deserving attention, are the *Custom House*, Third street; the *Court House*, Fourth street; *Temple of Justice*, Clark avenue; *Arsenal; Merchants' Exchange*, Main street; *Masonic Hall*, Market street; and some of the innumerable Hospitals, Asylums, Educational Institutions, and Roman Catholic Convents. Among Churches, the *Catholic Cathedral*, Walnut street, takes the lead; followed by *St. George's*, (Epis.) Locust street; *First Presbyterian*, Fourteenth street; *Church of the Messiah*, (Unitarian) Oliver street, etc. Theatres: *De Bar's Opera House, Olymvic* and *Varieties*. Prominent Hotels: the *Planters', Southern, Laclede, Everett*, etc.

Two other Objects of Interest at St. Louis demand special notice: the *Steel Bridge*, now building and nearly or quite completed across the Mississippi, from Washington Avenue to the Illinois shore, for railroad and general use, and undoubtedly destined to be one of the world's master-works in bridge-erection; and the *City Water Works*, not long completed, with tower, and elaborate machinery for straining and purifying the river-water, believed to be among the best in use.

[The transit connections of St. Louis, by railway and steamboat, are among the most extensive on the continent. By rail, east to *Indianapolis* by the St. Louis and Terre Haute road, and to *Cincinnati* and eastward by the Ohio and Mississippi road; southeastward to *Tennessee Cities* by the route just traversed; south to *New Orleans* and the gulf by the Mobile and Ohio road: west to *Jefferson City*, to *Topeka*, and other towns of Kansas, by the Pacific and Missouri road; northwestward to *St. Joseph, Omaha* and the Pacific Railroad for *Salt Lake City* and San Francisco, by the same and St. Joseph roads; north to *Chicago* and the Lakes and Canada, by the Chicago, Alton and St. Louis road. In addition, steamboat communication to all navigable points on the Mississippi, the Missouri and Ohio rivers.]

Leave St. Louis by Chicago, Alton and St. Louis road; by ALTON, loftily located at just above the junction of the Missouri and Mississippi Rivers, with grand and notable scenery at that point, and much general charm in situation [connection eastward to *Indianapolis* by the Indianapolis and St. Louis road]; to

SPRINGFIELD, capital of the State of Illinois, a thriving and handsome town standing on the margin of a wide and fine *prairie*. It has a very handsome *Capitol, State Arsenal, Court House, Custom House*, etc.; extensive *Water Works* on the Sangamon River; and will always enjoy an additional celebrity as the residence and burial place of *Abraham Lincoln*, a noble Monument to whom marks his tomb in *Ridge Cemetery*. Hotel: the *Leland House*. [Connection westward to *Quincy*, and eastward to *Logansport* and *Fort Wayne*, by the Toledo, Wabash and Western road.]

Springfield to BLOOMINGTON, capital of McLean county, and a large town of much commerce and many manufactures, besides having the great engine-shops of the Chicago and Alton Company. [Connection southwestward to *Jacksonville*; west to *Pekin*; southeast to *Champaign* and the Illinois Central road.] To *Chenoa* [connection west to *Peoria*; east to *Warsaw* and *Logansport*]. By other stations to JOLIET, large and thriving town on the Des Moines river, with State Penitentiary of noble construction, immense fine building-stone quarries near, valuable water-power, and extensive trade and manufactures. [Connection west to *Rock Island*, by the Chicago, Rock Island and Pacific road.] Joliet to CHICAGO.

(For notes on Chicago, see Route No. 13, Division C.)

ROUTE NO. 15.—NORTHERN AND WESTERN, (SEMI-SKELETON.)

BUFFALO TO CLEVELAND AND CHICAGO, BY LAKE SHORE ROAD.

Leave Buffalo by Lake Shore road; to *Dunkirk*, on Lake Erie, terminus of lower branch of the Erie road [connection eastward to *Salamanca*, for the Oil Regions.] Dunkirk to

ERIE, (Pa.), on the shore of Lake Erie, with fine harbor (a U. S. Naval Station); *Court House* and other good buildings; extensive iron rolling-mills, and the connection of the Erie Extension Canal with the Ohio River and Beaver Canal. Hotel: the *Reid House*. [Connection southeast to *Corry* and the Oil Regions]. By *Girard* [connection southward to Pittsburg]; and Painesville. to

CLEVELAND, (Ohio) on Lake Erie, with harbor at mouth of Cuyahoga River, heavy lake shipping trade, much prosperity in business aspects; and so pronounced a shaded beauty, especially in the fine elms lining its wide streets, that it bears the name of the "Forest City." It has a *Medical College:* a *Marine Hospital;* several handsome Churches; a splendid *Union Railway Depot*, of great size; *Monumental Park* (with Monument to Commodore Perry); *Woodlawn Cemetery;* noble *Water Works;* and many other attractions. [Connections southeast to

Pittsburg and *Wheeling;* south to *Coshocton* and *Zanesville;* southwest to *Columbus, Cincinnati,* etc.]

From Cleveland, by *Oberlin* (seat of the celebrated "Oberlin College," which admits blacks as well as whites); by *Monroeville* [connection northward to Sandusky]; by *Clyde* [connection south to CINCINNATI]; *Fremont,* etc., to

TOLEDO, on the Maumee River, near Lake Erie, with considerable lake trade (principally in grain), much domestic commerce, many handsome buildings, rapid progress, great educational facilities, and an almost matchless location as a railway centre. Hotels: *Oliver House, Island House, American,* and *St. Charles.* [Conections: southeast to *Clyde* and (opening) to Wheeling; South to *Lima, Dayton* and CINCINNATI; southwest to *Logansport, Springfield* (Ill), the Mississippi river, and *St. Louis;* northward to *Detroit* and the Canadian lines; etc.]

From Toledo, by the Michigan Southern and Northern Indiana road; by ADRIEN (Michigan), with water-power; some manufactures; repair-shops of the railroad-division; a handsome Soldiers' Monument, and many attractions as a residence. Hotel: the *Lawrence House.* By *Hillsdale; Jonesville* [connection south to *Fort Wayne*]; *Sturgis* [connection north to *Grand Haven* and south to *Fort Wayne*]; *White Pigeon* [connection north to *Kalamazoo*]; *Elkhart* [junction with Air-Line of same road, to TOLEDO]; *South Bend* and *Laporte;* to

CHICAGO. (For notes on· Chicago, and connections, see previous route, No. 13.)

ROUTE NO. 16.—NORTHERN (SEMI-SKELETON).

NEW YORK OR PHILADELPHIA TO THE LACKA-
WANNA COAL REGIONS, AND THE OIL CREEK
OIL REGIONS.

New York, by the New Jersey Central road to *Hampton Junction* (see Route No. 13); to

Manunkachunk (New Jersey).

Or by the Delaware, Lackawanna and Western road—Morris and Essex Division from New York: from foot Barclay Street to Hoboken; thence by rail, by *Orange* [connection to NEWARK]; by *Madison*, location of Drew Theological Seminary; by MORRISTOWN, thriving town of New Jersey, and capital of Morris County, on the Whippany Creek, with handsome residences, a "Washington's Head Quarters" and other Revolutionary remains; by *Boonton*, *Rockaway* and *Dover*, all towns in the iron-region, with extensive iron mills and foundries; by *Chester*, Drakesville, *Stanhope* [connection by stage or boat to *Lake Hopatcong* and to *Budd's Lake*]; by *Waterloo* [connection north to *Newton*, by Sussex road]; by *Hackettstown*, handsome town of Warren County, with flouring mills and a Methodist Episcopal Seminary of eminence [connection by stage to *Schooley's Mountain* (see Excursions from New York)]; to *Washington*. At Washington connect

ROUTE NO. 16.—NORTHERN.

with the Delaware, Lackawanna and Western road, and by that road to *Manunkachunk*.

Or, from Philadelphia, by the Northern Pennsylvania road, or the Belvidere Delaware road, to *Easton;* thence to *Manunkachunk*.

Manunkachunk to the *Delaware Water-Gap* (see Excursions from New York). Water-Gap to *Stroudsburg*, (Pa.), and by several minor stations to

SCRANTON (Pa.), important heavy-manufacturing town, and great centre of the coal operations of the Lackawanna district. In brief excursions from Scranton, locally directed, may be observed all the details of mining and transportation, of the immense coal trade and the iron trade accompanying. [Connections southwest to *Pittston* and *Wilkesbarre*; north to *Great Bend* and the Erie road; east to *Carbondale, Honesdale,* etc]. Scranton to

Pittston, another important coal centre. Pittston, by *Rupert* [connection southward with the wildly-grand Catawissa road, for *Reading*, etc.]; by *Milton* and other stations, to

WILLIAMSPORT, capital of Lycoming County, on the Susquehanna river, with much industry and miscellaneous business, and the most extensive lumber trade of any town in America. Among the curiosities of the place, is the *Great Timber Boom* in the Susquehanna, capable of holding millions of logs at a time; the many saw-mills and other lumber works. There are also extensive Black Marble

Quarries in the neighborhood. Hotels: the *Herdic, City,* and *American.*

From Williamsport by the Philadelphia and Erie road; by

LOCK HAVEN, another great lumber centre, also with immense Timber-Boom, saw-mills, etc., and fine scenery in the neighborhood. Hotels: the *Fulton, Irving,* and *Montour.* Lock Haven, by *Renovo, Emporium,* Wilcox and other stations; to

Irvineton, whence should be taken the Oil Creek and Alleghany road, to *Tidioute, Oleopolis, Pithole, Oil City, Titusville,* or any of those great oil centres, from which short excursions, locally directed, can be made with most profit and satisfaction. Thence to CORRY, for *Salamanca* and the Erie road, going east; or for *Erie* and the Lake Shore road, for the north or West.

ROUTE NO. 17.—CANADIAN AND WESTERN.

NIAGARA FALLS, BY HAMILTON AND LONDON TO DETROIT AND CHICAGO, BY GREAT WESTERN AND MICHIGAN RAILWAYS.

Leave Niagara Falls (Suspension Bridge), by rail on the Great Western road of Canada; by *Thorold* (crossing of the Welland Canal around the Falls of Niagara), to

St. Catharine's, pleasant small town, favorite as a residence, and with Mineral Springs of much celebrity. Thence by *Grimsby*, lying near the shore of Lake Ontario, to

HAMILTON, on Burlington Bay, at the extreme western end of the Lake, with very handsome coast-scenery in the neighborhood; a magnificent harbor, with heavy lake trade and fine fishing; and the town itself very prettily laid out, with elegant residences and other buildings, well shaded and attractive. Very fine views are to be obtained from the *Mountain*, where also stands *Dundrum Castle*, erected by Sir Allan McNab, when Governor-General; and there are many favorite resorts in the neighborhood, among others the *Beach, Oaklands, Flamborough Heights*, etc. Prominent Hotels: the *Anglo-American* and *City*. [Connection eastward (*Hamilton Junction*) with the Grand Trunk Railway for *Toronto* and all the Eastern Canadian cities; also, by boat on the Lake, for *Toronto* and the same]. Hamilton to DUNDAS, with many manufactures and much fine scenery at and near the Desjardines Canal, here commencing; to *Harrisburg* [connection northward to *Berlin, Guelph*, and the Grand Trunk road]; to

PARIS, a thriving town, with important water-power and manufactures, at the junction of the Grand and Nith rivers, with mineral springs and a petrifying spring in the neighborhood [connection northwestward to *Goderich* and Lake Huron, and southeastward to *Dunville* and BUFFALO, by Goderich and Grand Trunk road]. Paris to

LONDON, considered the metropolis of South Western Canada—with handsome location, streets well laid out and shaded, costly buildings, and all the attractions for residence. [Connection northward to *St. Mary's* and the Grand Trunk road; southward to *Port Stanley*, on Lake Erie, with boat connection to BUFFALO]. London, by *Komoka* [connection westward to *Petrolia* (oil-centre), and *Port Sarnia*, at the entrance of the St. Clair river into Lake Huron]; and by Glencoe; to

Bothwell, principal town of the Canadian oil-regions, in the neighborhood of which those who have not visited the Pennsylvania oil-sections, may derive a very good idea of the petroleum wells and processes.

Bothwell to *Chatham*, with the distinction of very

large percentage of negro population. [Connection with *Detroit*, by steamers down the Thames river and across Lake St. Clair to Detroit river]. Chatham by unimportant stations to

WINDSOR, very old town on the eastern side of the Detroit river, with many French peculiarities and but moderate prosperity. From Windsor, ferry, carrying over cars on boats, to

DETROIT, Michigan, lying on the west bank of the Detroit river, strait connecting Lake St. Clair with Lake Erie. This is the largest city of the State; one of the oldest in any of the Western States, and one of the most wealthy and influential of all. The city front extends along the river at great length, with most of the location elevated, and the streets well shaded and broken up into many small parks and public grounds. The most important of the latter is the *Grand Circus*, park and promenade, from which radiate many of the finest avenues; among others, *Woodward*, *Jefferson*, etc. There is also a large Plaza, called the *Campus Martius*, around which are grouped many of the finest buildings in the city. Without the town, the favorite public resorts are *Fort Wayne*, on the river, three miles from the city; *Belle Ile*, *Grosse Pointe* and *Grosse Ile*, more distant. The principal Cemeteries are *Elmwood* and *Woodlawn*.

Among the prominent buildings in the city, are the *Michigan Central Freight Depot*, of immense size and costly construction, with the great *Loco-*

motive Round House and *Grain Elevator*, near it; the *Custom House* (with Post Office); the *Opera House;* the *Board of Trade Building,* etc. The most notable churches are *St. Paul's* (Epis.), with the peculiarity of a roof without columns; *Christ, St. John's* and *Grace* (all Epis.); *Fort Street Presbyterian; Central* (Meth.-Epis.); *St. Peter* and *St. Paul* (Cath.); *St. Anne's* (Cath.), with very fine choir; etc. Detroit has also elaborate Water-works; large manufacturing and lake-shipping interests, in grain and provisions, etc. Theatre: the *Opera House*. Prominent Hotels: The *Russell, Biddle,* and *Michigan Exchange.*

From Detroit may be visited, north, *Lake St. Clair,* with many attractive features in scenery; and southward, the *Put-in-Bay Islands,* below the mouth of the Detroit river, in Lake Erie, near which occurred Commodore Perry's victory in 1813; now famous as bathing and fishing resort, etc. [Boat from Detroit to Kelly's Island, largest of the group, every day during warm season.]

[Railway connection from Detroit: north to *Port Huron,* foot of Lake Huron: northwest to *Saginaw, Wenona,* etc.; west to *Kalamazoo,* etc.; east by route just traversed, and by Toledo and Lake Shore road; west to CHICAGO, etc., as see route to be pursued. In addition, it has steamboat communication on Lake Erie to BUFFALO and other ports; and to all ports on Lake Huron and Michigan.]

Leave Detroit by Michigan Central road, by *Yp-*

silanti, pleasant small town of Michigan, seat of the State Normal School; thence (along the Huron River), to

ANN ARBOR, handsome large town, on elevated plateau, with fine shade and many handsome buildings; and seat of the *University of Michigan*, an institution of wide influence, with varied courses, a fine Observatory, etc. Hotel: the *Gregory House*. By *Dexter* and *Chelsea* to

JACKSON, large and thriving town, with many manufactures, an important coal-trade from mines in the immediate neighborhood, and seat of the *Michigan State Prison*. [Connections, northward to LANSING (capital of the State); southward to *Adrian* and *Toledo*, etc.] From Lansing, by *Parma*, on the Kalamazoo river; by *Albion* (seat of Albion College, of the Meth. Epis. Church); by *Marshall* (large paper manufactories, and railway repair-shops); to BATTLE CREEK, manufacturing town, especially with extensive flour-mills [connections north to LANSING, and south to *South Bend*, etc.]; to

KALAMAZOO, largest town in the State, after Detroit. It is a thriving manufacturing and commercial town, with much shaded beauty and many handsome residences; and the seat of a *Baptist College* and the *State Insane Asylum*. Hotels: the *Kalamazoo* and *Burdick*. [Connections: northwest to *Grand Haven* and Lake Michigan; also to *Grand Rapids*; southeastward to *Fort Wayne*, etc.] From

Kalamazoo, by *Lawton* (with extensive iron works); *Niles*, small town of commercial and industrial importance on the St. Joseph River; *New Buffalo* and *Michigan City* (both modern towns, on the immediate shore of Lake Michigan); to

CHICAGO.

ROUTE NO. 18.—NORTH-WESTERN (SEMI-SKELETON.)

CHICAGO TO ST. PAUL (MINN.) AND FALLS OF ST. ANTHONY; WITH OPTIONAL RETURN DOWN THE MISSISSIPPI OR BY LAKE SUPERIOR.

Leave Chicago by the Milwaukie Division of the Chicago and Northwestern road; by *Waukegan;* KENOSHA [connection west to *Genoa, Rockford,* etc.]; RACINE [connection west to *Elkhorn* and *Freeport*; to MILWAUKIE. (Or, by daily steamer on the Lake, direct from CHICAGO to MILWAUKIE.)

MILWAUKIE, commercial capital of the State of Wisconsin, one of the largest cities of the northwest, considered very handsome and attractive as a residence, and so healthful in reputation, as to have originated the jest that "people are obliged to go away from Milwaukie, when they wish to die!" Hotels: the *Plankinton, Walker,* and *Newhall.* [Connections, northwest to *Horicon, Portage City,* etc.; southwest to *Milton, Janesville,* etc.]

Leave Milwaukie by the Milwaukie and St. Paul road; by *Watertown* [connection north to *Horicon;* northwest to Portage City, etc.]; to

MADISON, capital of the State, and a very thriving and handsome town, with the *Capitol, University of Wisconsin,* many other local attractions, and the notably-beautiful *Four Lakes* in the immediate neighborhood. [Connections southeast to CHICAGO, by the Chicago and North-western road; and to *Plymouth, Beloit*, etc., by the Madison division of the same road]. Madison, by many minor stations, to

PRAIRIE DU CHIEN, important town on the Mississippi river, with a considerable river-trade, many steamboats making stoppage, and *prairies* in the neighborhood, as the name indicates. By ferry to

McGregor, small town on the opposite side of the river; where the route by rail is continued. By *Colmar* [connection westward to *Charles City* and the Missouri river]; to AUSTIN [connection southward by the Burlington and Cedar Rapids road, to *Cedar Rapids, Burlington*, and the Mississippi]; to *Ramsey* [connection west with Southern Minnesota road]; to *Owatona* [connection west for *Mankato, St. Peter*, etc.]; by *St. Paul* and *Mendota Junctions* [connection southwest to *Mankato*,] etc.; by *Minnehaha* and *Minneapolis;* to

ST. PAUL, capital of Minnesota, and the largest town in the State; on the Mississippi river, at the virtual head of navigation; with *State Capitol; State Reform School; St. Joseph's Academy* (Catholic); a *Bridge* of great length, over the Mississippi; *Carver's Cave* and *Fountain Cave* in the immediate vicinity, etc. Theatre: the *Opera House*. Leading Hotel: the *Merchants'*. [Connections: north to *Duluth*, on Lake Superior; northwest to *St. Cloud;* west to *Breckenridge;* southwest to *St. Peter* and *Mankato;* southeast to *Milwaukie* and CHICAGO, by route just traversed; also southeast to *Red Wing,* and Lake Pepin. Also by steamer to all Mississippi ports, ST. LOUIS and NEW ORLEANS.] It is from St. Paul that visit will be paid (short ride by carriage, by *Fort Snelling*) to the

Falls of Minnehaha, very beautiful small fall of the Minnesota river, made famous by Longfellow in the poem of the same name, with the Indian derivation, "Laughing Water." Also will be visited, by rail from St. Paul, the

Falls of St. Anthony, and town of the same name, ST. ANTHONY, a few miles above. The town is a thriving one, at the actual head of navigation of the Mississippi, with *State University,* and connection by bridge with *Minneapolis.* The Falls, though with very mean surroundings, are grand, especially in the feature of Rapids, and show to best advantage by *moonlight.*

From St. Paul descent of the Mississippi may be made, by steamboat, by *Red Wing* (Minn.); *La Crosse* (Wis.); *Prairie du Chien* (Wis.); *Dubuque* (Iowa); *Galena* (Ill.), centre of the lead trade; *Davenport* (Iowa); *Rock Island* (Ill.); *Burlington* (Iowa); *Nauvoo* (Ill.), original seat of the Mormons; *Keokuk* (Iowa); *Hannibal* (Mo.); *Alton* (Ill.), and many other interesting river ports, with stoppages, to ST. LOUIS for the South or return eastward.

Or, northern route may be taken, leaving St. Paul by the Lake Superior and Mississippi road, to

DULUTH, new but important town at the extreme southwest point of Lake Superior, with good harbor, heavy lake trade and rapidly increasing prosperity, Hotel: the *Clark House*. At Duluth, steamer to be taken (depending on local direction for the most reliable particulars) on Lake Superior, to the *Ontonagon Copper Region*, on the south shore of that Lake; thence to the *Marquette Iron Region*, on the same shore; thence to the *Pictured Rocks*, also on the same shore. Thence route may be continued. through the *Sault St. Marie* (Strait) into *Lake Huron*, and to *Bay City* for rail to DETROIT; or to DETROIT by boat direct; or to *Goderich*, for return by rail through Canada; or through the St. Clair River and Lake, and the Detroit river, to *Lake Erie*, for *Toledo*, *Cleveland*, *Erie*, or BUFFALO, on that Lake.

ROUTE NO. 19.—CANADIAN.

NIAGARA FALLS TO TORONTO, OTTAWA, MONTREAL, QUEBEC, AND THE SAGUENAY RIVER; BY GRAND TRUNK RAILWAY, AND BOAT-CONNECTIONS.

Division A

NIAGARA FALLS TO TORONTO AND OTTAWA.

Leave Niagara Falls (Suspension Bridge) by Great Western Railway, by *Thorold*, to *St. Catharine's* (see Route No. 17); and to

HAMILTON (also see Route No 17.)

From Hamilton, continuing by Toronto branch of Great Western Railway, at near the upper coast of Lake Ontario, by *Oakville* and other stations, to

TORONTO, most populous city of the Western province (Ontario), and one of the handsomest in America, though excelled in size by many. It lies on the Northern shore of Lake Ontario; is well laid out and finely shaded; and has one thoroughfare, *Yonge Street*, actually extending northward as an unbroken drive, the whole distance to the foot of *Lake Simcoe*, some 35 miles. Among the prominent buildings is the *University of Toronto*, a noble structure with lofty tower, and fine park surrounding. Scarcely second is *Osgoode Hall*, the law-court building, with the distinction of not only being one of the most tasteful in the world for legal purposes, outside, but one of the most completely and tastefully arranged, within. There are also the *Exchange, Provincial Lunatic Asylum, Trinity College, Normal School*, etc., all worthy of visit. Of the many Churches, three have especial prominence: the *Cathedral of St. James* (Epis.); that of *St. Michael* (Catholic); and the (new) *Wesleyan Church*. Prominent Hotels: the *Rossin House*, and the *Queen's*. Those who have abundant leisure, should make the drive before spoken of, to *Lake Simcoe*, with wild beauty; those with less time will find drives through some of the main avenues, and along the shore of the Lake, amply repaying them.

[Connections by rail, north to Lake Simcoe; west to *Guelph, Berlin*, and other towns on the Grand Trunk road. Also, by boat with *Niagara Falls*, by *Lewiston* and rail along the Niagara River. Also, by daily boat along Lake Ontario and down the St. Lawrence River to MONTREAL. Also, to ports on the New York side of the Lake.]

From Toronto by the Grand Trunk Railway; by *Frenchman's Bay, Bowmanville* and other stations, to

Port Hope, pleasant little town, on the Lake, with hill suburbs and some lake-trade. [Connection northwestward to *Beaverton*, on Lake Simcoe; and with Lake ports, by boat.] Port Hope to

Cobourg, important station as well as handsome

town, with fair trade, a pleasant residence, and the seat of *Victoria College* (Wesleyan). [Connection northward to *Peterboro* and Rice and Salmon Trout Lakes.] Cobourg, by *Colborne* (not to be confounded with "Port Colbourn," on the Great Western road); by *Trenton* (on the little river Trent); by *Belleville*, pleasant small town on Moira river (actual inlet from the Lake); by *Napanee, Collins' Bay* and other stations; to

KINGSTON, very old town at the entrance of the St. Lawrence river, and once capital of Canada; with very heavy fortifications, in *Fort Henry* and several other works; seat of *Queen's College University*, the *Regiopolis Catholic College*, Provincial Penitentiary. Hotel: the *British American*. *Cape Vincent*, on the New York shore, lies opposite. [Steamer connection from Kingston up the Lake to Toronto, down the Lake to MONTREAL and other ports. Also, by Rideau Canal, with *Ottawa*. From Cape Vincent, by rail to *Watertown*, and thence to *Rome* and other points on the New York Central road.]

From Kingston, by *Gananoque* and *Mallory Town*, to BROCKVILLE, a town of pleasant location and some commercial importance, on the St. Lawrence river [connection northward to *Carleton Place, Arnprior*, etc.; and from Carleton Place, by Canada Central road, to OTTAWA]. Brockville to

PRESCOTT (Junction), small town, principally of transit importance, also on the St. Lawrence. [Connections, north to Ottawa, as see route to be pursued; east direct to Montreal, continuing by Grand Trunk road; across the river to *Ogdensburg*, on the New York side, whence connection south to the New York Central road and towns on that line, by the Rome, Watertown and Ogdensburg road, for southward; or eastward by Ogdensburg and Champlain road to *Rouse's Point*, Lake Champlain and *St. Albans* for all Eastern States].

From Prescott, by St. Lawrence and Ottawa road, through a region comparatively unbroken, but with many features of beauty in natural scenery, to

OTTAWA, Capital of the Dominion of Canada. It lies on the Ottawa river, and on the Rideau Canal, running to Lake Ontario at Kingston. The scenery in the district is somewhat wild and untamed, but very picturesque; and in the immediate neighborhood, at the *Falls of the Rideau*, may be witnessed with interest some of the heavy lumbering operations, in the *timber-shoots* down the inclined planes to avoid the Falls. The *Chaudiere Great Falls* (of the Ottawa), lie within the city proper, at the west, are some 200 feet in width by 40 in depth, and have many features of grandeur; while the *Little Falls*, handsomer though smaller, lie at the east. The *Rideau Falls*, at the northeast, and the *Remoux* and *De Cheyne Rapids*, some miles above, are all worthy of visit and notice.

The feature of Ottawa is of course to be found in the *Parliament Houses* and government buildings connected. They are of native stone, lately erected, at great cost, and truly magnificent in size, design and arrangement—promising, when fully completed, with their grounds, to be worthy of the Dominion and command great admiration. The respective *Chambers* of the Senate and House of Commons are of the same size as those at Westminster Palace, and quite as handsomely finished; and there is an immense *Library*, not yet entirely filled, capable of accommodating half a million volumes. They stand at the height of an elevation known as "Barrack Hill," forming the apex of the higher ground on which the Upper Town is built, being divided from the Lower Town by the Rideau Canal and its handsome stone bridge. The *Queen's Printing House*, near the Parliament Houses, the *Catholic Cathedral*, and other prominent buildings, demand attention. Leading Hotel: the *Russell House*. [Connection west to *Carleton Place* and the Brockville road; and to *Prescott, Ogdensburg*, etc., by the route just traversed. Direct railway along the Ottawa river, to MONTREAL, in course of construction].

Division B.

OTTAWA TO AND AT MONTREAL.

Leave Ottawa by morning boat on the Ottawa river; with fine view, soon after leaving, of the *Rideau Falls*, on the right; and not long afterwards, the entrance of the *Gatineau River*, tributary of the Ottawa, into that river, the banks of which, almost from the first, are rough, wooded and picturesque. Landings are made at *Buckingham*, at *Thurso* (flourishing village, with a large lumber trade), at *Brown's* and *Major's*; before reaching

L'ORIGINAL; at which point, if time allows, the tourist should lie over for one day, to visit the

Caledonia Springs, nine miles distant, the healing qualities of which have made them very celebrated, with capacious hotel, and the presence, in summer, of a very large number of the best known people of fashion and condition, in the Dominion. Return to *L'Original.*

From L'Original, whether with or without having visited the Springs, the course is pursued by boat, down the Ottawa, to

GRENVILLE (with *Hawkesbury* opposite, with large saw mills) where the boat is left and a land ride of twelve miles taken (the *Long Sault Rapids* making navigation impossible); to

CARILLON, at the lower end of the rapids, where another boat of the line is taken. It is worthy of remark that from Ottawa to this point, the middle of the river has been the dividing line between the two provinces of Ontario (west) and Quebec (east); but that here the line leaves the river, striking southward to the St. Lawrence, and the course is taken entirely in that of Quebec.

Among the next prominent objects of interest following, is the *Mountain of Rigaud*, looming high on the southern bank, above the rough and wooded shores. Landings are made at *Pointe-aux-Anglais*, *Hudson*, and *Como;* after leaving the latter of which the river expands into the

Lake of the Two Mountains, with the two mountains giving its name, rising on either side, one of them, *Calvary*, being held sacred by the Indians. Not long after, is reached

ST. ANNE'S, rendered so celebrated by Moore, in the "Canadian Boat Song." At this point *Mont Royale*, the height above Montreal, comes into view. The boat is taken through locks, at St. Anne's, to avoid the rapids, coming out into the

Lake St. Louis, in which the Ottawa for the first time joins the St. Lawrence. Landing from the boat is made at

LACHINE, where cars of the Grand Trunk Railway are taken to

MONTREAL, the largest and most prosperous city of the British Possessions in North America; Metropolitan See of the English Church in Canada, and seat of a Catholic Bishopric. It lies on rapidly rising ground, on the island of the same name, with the St. Lawrence immediately in front, Back River forming the sound behind it; and the mountain which gives it name, *Mount Royal*, also rising grandly at the back. There are few and unimportant fortifications, (*St. Helen's Island* being the principal); but the garrison generally kept is large, the place being considered the military key of the

Dominion. The population of Montreal is wondrously mixed, there being many streets, in the higher and newer parts of the town, in which the English and Scotch elements entirely predominate, with many of the features of an English city; while in the older and lower parts of the town, many of the streets are still called "Rues," and the prevailing architecture, language and manner are all French, of not too refined an order. There are now fine quays along the river; costly and elegant residences have rapidly increased in number, stretching back towards Mount Royal; and the commercial importance and prosperity of the Northern Metropolis have quite kept pace with its growing luxury. Among other evidences of its prosperity has been the establishment of the fine *Allan line of Steamers* to Liverpool and Glasgow, coming to Quebec and Montreal during the open season, and to Portland in the winter.

First among the edifices of Montreal, comes the *Catholic Cathedral of Notre Dame*, standing on the *Place d'Armes*, in the centre of the old city, and so large that it is accredited with containing 10,000 people without difficulty. It is Gothic in architecture, with two tall towers, commanding a magnificent view from the top; and within, it has many of the features of European churches of the same faith. *Christ Church Cathedral* (Epis.) and *St. Andrew's Church* (Epis.) rank next, commanding much admiration; and there are many other churches, more or less creditable in taste. Of other buildings may be especially noted the *Court House*, one of the best on the continent; the *Bank of Montreal*, near the Cathedral; *McGill College*, at the foot of Mount Royal; *Bonsecours Market*, on the quay, with large dome and excellent internal arrangements; *St. Patrick's Hall*, Victoria square; the *Albert Buildings*, same place; *Dominion Block*, McGill street, etc. There are three *Nunneries*, always exciting more or less attention among visitors, and to which admission is often granted; the *Gray*, Foundling street; the *Black*, Notre Dame street; and the *Hotel Dieu*. *Notre Dame* and *Great St. James Streets* may be named as the most fashionable promenades; and *St. Paul Street* as the leading commercial. The principal Cemetery is *Mount Royal*, on the mountain of that name, around which, also, is the most fashionable drive of the city. There is a handsome *Nelson Monument* at the Place Jacques Cartier. Theatre: the *Montreal*. Leading Hotels: the *Ottawa*, *St. Lawrence Hall*, *Donegana's*, the *Montreal*, etc.

Many excursions of interest can be made from Montreal, but the most indispensable one is that to

The Victoria Bridge over the St. Lawrence, at Point St. Charles, order to inspect which can be obtained from the officers of the Grand Trunk Railway, near the entrance. It is one of the immense enterprises of later times, with no less than 23 spans of 242 feet each, a centre one of 330 feet, and a total length of two miles. It is tubular, on the plan of the great bridge over the Menai Strait, in Wales; was built by Robert Stephenson and A. M. Ross; and opened by the Prince of Wales during his American visit, in August, 1860.

Another very pleasant excursion, for those reaching Montreal by rail, is to take rail to *Lachine*, and thence return to the city by boat *Down the Lachine Rapids*, with excellent idea thus obtained of that feature of the St. Lawrence.

[Connections from Montreal: westward to OTTAWA, by the route just traversed; eastward to QUEBEC, by the Grand Trunk road (from St. Lambert); southward to *Rouse's Point*, and thence to all points in the Eastern and Middle United States, (from La Prairie). Westward by steamer on the St. Lawrence and Lake Ontario, to *Toronto* and leading Lake ports; eastward by steamer on the St. Lawrence, to QUEBEC and the farther East. By Allan line of sea-steamers to PORTLAND; and to HALIFAX, LIVERPOOL and GLASGOW.]

Division C.

MONTREAL TO AND AT QUEBEC, WITH EXCURSIONS.

Leave Montreal (from Bonaventure Station), by train on the Grand Trunk Railway; across the *Victoria Bridge* at *Point St. Charles;* by *St. Lambert, St. Hyacinthe*, and other stations, to

RICHMOND, important railway station and point of intersection. [Connection, south-eastward, by Portland Division of the Grand Trunk road, to *Island Pond*; and thence to *Gorham* for the White Mountains and southward, or to PORTLAND and connections for BOSTON and the east.]

From Richmond, by the Quebec branch of the Grand Trunk road; by ARTHABASCA [connection northward, by *Bulstrode*, to *St. Gregoire*, on the St. Lawrence river, and *Three Rivers* (Canada)]; by *Black River* and other stations, to

Point Levi [continuation of line eastward, to *Riviere du Loup*]. From Point Levi, ferry across the St. Lawrence, to

QUEBEC, metropolis of the Province of the same name; important military station, with very strong fortifications; and with as important historical interest as any city on the American continent. It lies on the north shore of the St. Lawrence, with exceedingly picturesque location, being divided into the Upper and Lower Towns, with the very strong fortifications of the Upper Town crowning the whole; and the *Citadel of Cape Diamond*, being considered next in strength in the world to Gibraltar and Ehrenbreitstein. From the city proper, the suburbs of St. Roch and St. John extend along the river St. Charles to the *Plains of Abraham*, on the *Heights* of the same name, rendered ever-memorable by the battle fought there between the English Gen. Wolfe and the French General Montcalm, in 1759, with the death of both the commanders, but the total defeat of the French, and the final capture of Quebec and destruction of the French power in the province. The spot where Wolfe fell, near an old redoubt at the highest point, is pointed out to tourists, who have even a more singular interest in seeing the skull of Montcalm, exhumed not many years ago, now preserved in the Ursuline Convent. The joint *Monument to Wolfe and Montcalm* is to be found in the *Public Garden,* on Des Carrieres street. At the foot of the Citadel is a tower, where the American General Montgomery fell in the assault on Quebec, in 1775. Ascent from the Lower to the Upper Town is made by a very steep and winding street, through the *Prescott Gate,* by which also the fortifications may be reached on the St. Lawrence side. The Plains may be entered by the *St. Louis Gate,* nearly opposite. The *View from the Citadel,* over the city, the St. Lawrence and the opposite shore, is a truly magnificent one and not to be omitted by any one with an eye to the picturesque.

Among the most notable Buildings of Quebec, may be named the *Parliament House* (rebuilt when the city was still expected to remain the Capital); the very large *Artillery Barracks;* the immense and fine *Roman Catholic Cathedral;* the *Ursuline Convent* and *Church,* with attractive gardens; the *English Cathedral* (modern and noble); *St. Andrew's Church;* the very old church of *Notre Dame des Victoires,* in the Lower Town; as also, in the Lower

Town, the *Exchange, Custom House, Marine Hospital, Post Office,* and many of the most extensive commercial establishments. It is worthy of remark that Quebec, even more than Montreal, has a large French admixture, and that in some of the quarters many of the lower Parisian dwellings and habits may be seen duplicated. Among the principal streets are *St. Louis; D'Auteuil* (near the Esplanade, with many fine residences); *St. Louis Road* (from the Gate of the same name); *St. Peter* (Lower Town)—commercial. Principal Cemetery: *Mt. Hermon,* elevated and handsome. Leading Hotels: the *St. Louis* and *Russell House.*

[Connections: southwestward to *Richmond* and MONTREAL, by route just traversed; southward by the same route, by Richmond, to *Island Pond,* PORTLAND, the White Mountains, etc.; eastward to the *Riviere du Loup.* Also, by boat on the St. Lawrence, to MONTREAL, etc.]

Of short Excursions from Quebec, the most important are those to the FALLS OF MONTMORENCI, noble broken cascade, with fine surrounding scenery, reached in drive from the city, through *Beauport* (seat of the *Provincial Lunatic Asylum*); to *Lorette,* a famous Indian Village, very popular for tourists, and with a world of Indian goods for sale; to *Cape Rouge* ("Carouge"), with fine river scenery; to the *River* and *Falls of the Chaudiere,* below the city (by Point Levi); to *Lake St. Charles,* with fine scenery and good angling. Something longer is that to the

Falls of St. Anne, which may, however, be taken in connection with those of Montmorenci.

Division D.

QUEBEC TO RIVIERE DU LOUP AND THE SAGUENAY RIVER.

Leave Quebec by rail, by Point Levi, along the south shore of the St. Lawrence, by *Chaudiere Junction, St. Thomas, L'Islet* and other stations, to *Riviere du Loup.*

Or, better, if time will allow the additional day:

Leave Quebec by one of the steamers of the Canadian Navigation Company (usual trips twice a week: timely reference on this point to be made at the hotel of stoppage, at Quebec.) First object of interest, the large *Island of Orleans,* in the St. Lawrence immediately below the city, with considerable prosperity. The *Falls of St. Anne* (before referred to), and *Lake St. Charles,* celebrated for fine trout-fishing, are both passed, at some miles below, though of course not visible from the boat. The first landing is made at

MURRAY BAY, on the north shore, pleasant village and attractive watering-place, much resorted to by Canadian families, and with good accomodation. Going on by steamer, an hour and a half later is reached

RIVIERE DU LOUP, on the southern side of the now rapidly-widening St. Lawrence—terminus of the easternmost branch of the Grand Trunk road. [Connection by rail, southwest to *Chaudiere Junction*, for QUEBEC; or thence to *Richmond*, for MONTREAL, or south to *Island Pond* for PORTLAND or the White Mountains]. [From Riviere du Loup, visit is paid, by stage, to the favorite watering-place of

Cacouna, with fine bathing, fishing, and much fashionable resort.]

Leaving Riviere du Loup, again by steamer, the St. Lawrence is recrossed, to the entrance of the

SAGUENAY RIVER, with scenery of such grand and stupendous wildness as is seldom encountered on either continent; the almost perpendicular cliffs at many points, and the great height of the bordering hills, combining with the darkness of the water, the frequent waterfalls, and the general aspect of wild desolation, to awe as well as enrapture. At very near the entrance of the river is passed the very old village of *Tadoussac;* and not long after, the little cove containing a fishing station, called *L'Ance a l'Eau*. The next points of interest reached, are the two frightful over-hanging cliff-mountains, *Cape Eternity* and *Cape Trinity*, beneath which, from the apprehension that they may fall at any moment, the tourist has no wish to remain for any long period, while the water seems black as ink, from the shadows. *Statue Point* and the *Tableau* are other points of special interest, approaching *Ha Ha Bay*, a beautiful village, amid softened scenery, where the route terminates.

Return by steamer to *Riviere du Loup*, whence rail to QUEBEC, or to *Chaudiere Junction* for proceeding southward.

ROUTE NO. 20.—CANADIAN.

NIAGARA FALLS TO TORONTO, MONTREAL AND QUEBEC, BY STEAMERS; WITH DIRECTION TO OTHER CITIES OF BRITISH POSSESSIONS.

Leave Niagara Falls (Suspension Bridge), by train to *Lewiston*, small town on the American shore of the Niagara River (QUEENSTON, larger town, on the Canadian side, opposite. with monument to the British General Brock, killed there in battle in 1812).

At Lewiston take Toronto boat (twice daily), on the Niagara river, with stop at

NIAGARA (Village), place of embarkation for other passengers from Falls by rail on the American side. Immediately below Niagara are passed *Fort Niagara*, on American side, and *Fort Massasauga*, on Canadian. Soon after, passing from the River into Lake Ontario, with short sail to

TORONTO (See Route No 19).

At Toronto take Royal Mail steamer for Montreal (every day, in connection with boat and train). Several hours' sail along the Lake, with shore-views, principally north—to

KINGSTON. (See Route No. 19).

(Or, leave Niagara by rail, as in Route No. 19, to

Hamilton, Toronto and *Kingston*, first taking boat here, at very early morning or afternoon).

At Kingston the Lake narrows to become virtually the St. Lawrence River, though, still very wide, and called the "Lake of the Thousand Islands," as containing the celebrated

Thousand Islands, said to number nearly twice as many, and certainly studding the stream very thickly, in rough-tree-crowned, wild and picturesque beauty—there really seeming, at times, to be difficulty in finding passage between them.

Leaving the Lake of the Thousand Islands, entering the St. Lawrence proper, passing *Ogdensburg* on the American side, and *Prescott*, on the Canadian (connection to Ottawa—see Route No. 19),— are soon entered the

Rapids of the St. Lawrence, among the most extended and notable to be found in any river on the globe, and some of them startling to the inexperienced who mark the rapid rush of the water and the sharp inclination of the boat, at the worst moments; though the amount of danger involved, with good boats and the inevitable skilful pilotage, must be almost nothing, as accidents are literally unheard of. The different Rapids follow each other in the succession named: the *Gallopes* (4); the *Plate*; the *Depleau*; the *Long Sault*; the *Coteau*; the *Cedars* (considered by many the finest); the *Cascades;* and the *Lachine* (shortest of all, but more sensational than any of the others). Immediately after passing the Lachine, is in sight, and soon after reached,

MONTREAL. (See Route No. 19).

From Montreal (every evening) by boats of the Richelieu Company, making the whole passage during the night (little interesting scenery offering), and landing in the morning at

QUEBEC. (See Route No. 19; as also for excursion to Riviere du Loup and the Saguenay River).

SKELETON ROUTES TO OTHER TOWNS OF BRITISH POSSESSIONS.

HALIFAX, Capital of Nova Scotia. Reached by steamers of the Allan line, from Montreal or Portland; or by packet-steamer direct from Portland; or from St. John, N. B., by steamer to Windsor, N. S., and rail thence to Halifax. Hotels: the *Waverley*, *Stewart's*, *Halifax*, and *International*.

ST. JOHN, New Brunswick. From Boston, by steamer, twice a week. From Halifax, N. S., by rail and steamer, by Windsor, N. S. Hotels: *Waverley* and *Stubbs'*.

FREDERICTON, Capital of New Brunswick. From Boston, by steamer to St. John's, and small steamer up the St. John River. Hotel: the *Barker House*.

WINDSOR, Nova Scotia. By rail from Halifax.

SIDNEY, Cape Breton. By boat from Halifax.

SHEDIAC, New Brunswick. By rail from St. John, N. B.; also by steamer from Quebec.

CHARLOTTE TOWN, Prince Edward's Island. By

rail from St. John, N. B., to Shediac, N. B.; thence boat.

PICTOU, Nova Scotia. By rail from St. John, N. B., to Shediac, N. B.; thence boat.

BATHURST, New Brunswick. By boat from Shediac; also from Quebec.

ROUTE NO. 21. FAR-WESTERN (SEMI-SKELETON.)

CHICAGO TO OMAHA, SALT LAKE CITY, SAN FRANCISCO, BIG TREES AND YO-SEMITE VALLEY.

Division A.

CHICAGO TO OMAHA, BY OPTIONAL ROUTES.

By Chicago and North Western road.

Leave Chicago by Chicago and North Western road, to

Junction [connection north to *Milwaukie;* north to *Fort Howard* and *Green Bay;* north-westward to MADISON; westward to *Dunleith* and *Dubuque*]. Junction, by *Geneva* and other stations, to

Dixon [connection northward to *Freeport;* southward to *Bloomington,* SPRINGFIELD, *Alton* and ST. LOUIS]. By other stations to

Morrison [connection southwestward to Rock Island]; to

CLINTON, on the Mississippi River, entering the State of Iowa [river connections north and south]. Clinton, by various stations, to

CEDAR RAPIDS, railway centre on the Red Cedar River. [Connections, northeast to *Dubuque;* north to *Waterloo, Austin* and ST. PAUL; south to *Burlington* and *Keokuk*]. Cedar Rapids, by various other stations, to

Marshall [connections northward to *Mason City, Austin* and ST. PAUL; southward to *Ottumwa, Keokuk,* etc.]; to *Boone,* thriving town and coal centre, commencement of the Western Division of the road; to

GRAND JUNCTION [connection north to *Fort Dodge,* thence to *Sioux City;* south to DES MOINES, capital of the State of Iowa]. Grand Junction, by many other stations, through the Valley of the Des Moines, to

MISSOURI VALLEY JUNCTION. [Connections north to *Sioux City;* westward, by California Junction, across the Missouri river to *Fremont* and the Union Pacific road]. By other stations to

COUNCIL BLUFFS, on the eastern side of the Missouri river. [Connections north to *Sioux City,* etc.; south to *Nebraska City* (by branch), LINCOLN, Capital of Nebraska, ST. JOSEPH, on the Missouri river, etc]. From Council Bluffs, ferry to OMAHA.

By Chicago and Rock Island road.

Leave Chicago by the Chicago and Rock Island road; by *Englewood* [connections eastward to all cities on the Michigan Southern and Lake Shore roads; southeastward to *Fort Wayne, Pittsburg,* etc]. By other stations to

JOLIET, large town on the Des Moines river, with State Penitentiary, extensive stone-quarries, etc. [Connections, eastward to Michigan Southern and Lake Shore roads; southwestward to *Blooming-*

ton and *Springfield;* also by Canal with CHICAGO]. By other stations to

LA SALLE, flourishing town and coal centre on the Illinois river. [Connections, north to *Mendota* and *Freeport*, south to *Bloomington* and SPRINGFIELD, by Illinois Central road; also by steamer to ST. LOUIS]. La Salle, by *Bureau* [connection southward to *Peoria*]; by *Pond Creek* [connection southwestward to *Quincy* and to *Burlington*]; by minor stations to

ROCK ISLAND, important town on the Mississippi, with extensive manufactures and river trade. [Connections northeast to *Freeport*, etc.; southeast to *Peoria;* south to *Alton* and ST. LOUIS; also by steamboat to ST. LOUIS]. From Rock Island, by bridge over the Mississippi, to

DAVENPORT (Iowa), large town on the western bank of that river, with water-power, manufactures, *Griswold* and other *Colleges*, an *Opera House*, etc. [Connections, substantially same as Rock Island]. Davenport to *Wilton* [connection southwestward to *Muscatine*, *Washington*, and the Kansas Pacific road]; to *Moscow* [connection, by Ashland, with Des Moines Valley road]; to

West Liberty [connection south to BURLINGTON; north to *Cedar Rapids*, etc.]; to

Iowa City, on the Iowa river, formerly capital of the State, and now with State University, manufactures, etc. By other stations to *Grinnell*, seat of Iowa College [connections north to *Mason City*, etc.; south to *Ottumwa*, etc.] By other stations to

DES MOINES, capital of the State of Iowa, thriving manufacturing town and coal centre, at the confluence of Des Moines and Raccoon rivers, with magnificent State House in course of erection. [Connections, northwest to *Fort Dodge* and *Sioux City;* southeast to *Ottumwa, Keokuk,* etc.] Des Moines, by *Dexter, Casey, Atlantic,* and other stations, to

COUNCIL BLUFFS and OMAHA. (See Chicago and Northwestern route.)

Division B.

OMAHA TO OGDEN AND SALT LAKE CITY.

OMAHA, Nebraska, on the western side of the Missouri river, opposite Council Bluffs, well located, and unprecedentedly rapid in growth, though deriving its principal importance from the great Pacific transit through it, and the commercial supply of a wide section, making it the central point between Chicago and San Francisco. Communication with Council Bluffs by ferry boat, and by the magnificent iron bridge now approaching completion. [Connections: (besides the routes just traversed) southeast to *Ottumwa, Keokuk,* and BURLINGTON, by the Burlington and Missouri road; north to *California Junction* and *Sioux City;* south to ST. JOSEPH, *Wyandotte,* and TOPEKA, capital of the State of Kansas; etc.]

ROUTE NO. 21.—FAR WESTERN.

Leave Omaha by Union Pacific Railroad, nearly due westward across Nebraska; by many minor stations and the more interesting ones of *Gilmore* (entrance of the Papillon Valley); *Elkhorn* (crossing of the Elkhorn river, near, and entrance of the Platte Valley); *Fremont* [connection with the Chicago and Northwestern road, at *California Junction*]; *North Bend*, with first views of the sandy Platte River; *Schuyler*, with thriving colony of Nova Scotians; *Columbus* (with bridge across the Platte, and railroad crossing the Loup Fork river not far beyond); *Grand Island*, with German settlement and flouring-mills; *Kearney*, supply station for Fort Kearney, a few miles distant on the opposite side of the river; *Plum Creek*, scene of the "Plum Creek Massacre" of railway employés in 1868; *McPherson*, supply station for Fort McPherson, on opposite side of the river; *North Platte*, with machine shop of the railway company; *Alkali*, with remains of the once famous "Alkali Station" of the stage route; *Ogalalla*, near the old stage-crossing of the Platte; *Julesburg*, with Fort Sedgwick near and in sight; *Sidney*, largest station on the line, with railway-repair-shops, a small military post, etc.; *Pine Bluffs*, with singular rock scenery in the neighborhood; and *Hillsdale*, with first views, just beyond, of the snow-crowned Rocky Mountains, especially "Long's" and the "Spanish Peaks"; to

CHEYENNE, (Wyoming Territory), principal station between OMAHA and OGDEN, on ground of nearly 6,000 feet above the sea, with railway-shops and much industry.

[Connection, south to DENVER, Colorado, for the best views of the grand Colorado Mountains; or for *Central City, Golden City, Pike's Peak*, or other mining and mountain centres, reached by stage from Denver. Also, for *Santa Fe* and all points in New Mexico.]

From Cheyenne, on sharp up-grade, and with grand mountain views southward, and amid very wild scenery; by *Hazard, Granite Canon* and *Buford*, to

SHERMAN, highest railway station in the world, 8,235 feet, and with fine air and many scenic attractions. From Sherman, by *Fort Saunders*, near the Laramie River; by

LARAMIE, with railway machine-shops, near the North Park on the south and the Black Hills on the north, and very favorable for residence and mountain rambles; *Carbon*, with coal-mines in the neighborhood; *Percy*, with view of the Elk Mountain; *St. Mary's*, with especially wild and rugged scenery, and with another crossing of the Platte; to

CRESTON, at the summit of the dividing ridge of the continent. From Creston, by *Bitter Creek;* with repair-shops and the entrance to the Bitter Creek Valley; by *Green River*, fording-place of the old overland-stage line, and with fine views of the Uintah Mountains at the south and the Wind River Mountains at the north; by

BRYAN [connection by stage with the great *Sweet Water Mining Region* and the once-popular *South Pass* of the overland emigration]; by GRANGER (Utah, and entrance of the Territory); by *Carter*, supply station for Fort Bridger, lying near; by *Aspen*, highest point of the road over the Wasatch Mountains; *Wasatch*, with tunnels following; *Castle Rock*, with grand scenery, at the entrance of

Echo Canon, one of the wildest rocky defiles in the world, and intimately connected with Mormon history. Beyond are passed *Echo City*, on the Weber river; *Weber Canon*, only less grand than the Echo; then *Weber Station; Devil's Gate;* with the Weber river seen rushing through a narrow gorge; *Uintah*, to

OGDEN, terminus of the Union Pacific road, though with small other importance. Hotel: the *Ogden House.*

At Ogden detour is made, by the Utah Central Railway, to

SALT LAKE CITY, in the Valley of the same name, at the base of the Wasatch Mountains—home of the Mormon religion and ascendancy, and in many regards one of the world's wonders. Prominent buildings: the *Lion* and *Bee Hive Houses* of Brigham Young; the *Tabernacle*, with very large organ; the *Endowment House;* the *Temple* (mere commencement); the *Theatre;* the *City Hall*, etc. Hotels: the *Salt Lake, Townsend, and Revere.* Northeast of the city, elevated, is the *Cemetery*, entirely destitute of decoration. *Camp Douglas*, the U. S. Military Station, lies two miles east of the city. The *Great Salt Lake* (another "Dead Sea") may be visited in a brief excursion from the City; and return may be made thence to Ogden.

Return to Ogden for pursuing the route to California.

Division C.

OGDEN TO SACRAMENTO AND SAN FRANCISCO.

Leave Ogden by the Central Pacific Railway, by

Corinne, important station [connection northward by stage to *Virginia, Helena,* and other mining and mountain towns of Nevada. Also, carriage connection to Salt Lake.] From Corinne to

Promontory Point, spot where the "last spike" was driven and the concluding celebration of the Pacific Railway held, 10th May, 1869. Shortly after leaving Promontory, is entered upon, the

Great American Desert, with no vegetation, but alkali-dust and desolation. By *Kelton* [connection by stage to *Boise City, Walla Walla,* PORTLAND, and other places in Idaho and Oregon.]; by *Toano; Pequop,* at entrance of the Humboldt Valley; *Wells,* with the "Humboldt Wells" in the neighborhood, believed to be craters of extinct volcanoes. The Humboldt River and Valley are followed, to *Osino,* termination of the Valley; to

ELKO, Nevada, important station and county capital [connection by stage and wagon to *White*

Pine, Wyoming, and other mining districts.] By *Carlin,* another important station, and rival of Elko; by *Palisade,* with rocks in the neighborhood giving it name, and distributing trade to mining regions south; by *Argenta,* with distributing trade to Reese River and White Pine Mines, [connection by stage to *Austin* and *Belmont*]; by *Battle Mountain;* by *Winnemucca,* with railway shops and mining trade [connection by stages to *Boise City, Paradise, Silver City,* etc.] The next feature of importance is the

Great Nevada Sandy Desert. By many minor stations, to *Wadsworth,* with extensive workshops and the commencement of ascent of the Sierra Nevada Mountains, to

RENO, important station, on the Truckee River, and great mining centre of supplies and transportation. [Stage connection south to *Virginia City,* (the "Comstock" and other great silver lodes), *Gold Hill, Carson, Washoe,* etc.] From Reno by several other stations, to *Boca,* entrance of the State of California, with steep ascents and snow-sheds following. Next is reached

TRUCKEE, handsome town, with many saw-mills, and point of leaving the railway for *Lake Tahoe,* southward, and *Lake Donner,* near the town, both very beautiful mountain lakes, but the former considered by many the most beautiful in any land. From Truckee the mountain scenery is very grand, though with constant snow-shed interruptions; by

Summit Station, highest point of the road over the

Sierras; by *Colfax* [stage connections by *Grass Valley, Nevada*, etc., to *Downieville*]; by *Auburn* [stage connection to *Coloma, Placerville* and *Georgetown*]; by minor stations, to

JUNCTION [connection for Northern California and Oregon]; to

SACRAMENTO, on the Sacramento river, at the junction of the American Fork, Capital of the State of California, and provincial railway centre of the State. Owing to many fires and equally many inundations, it lies in a disorganized condition, but has many objects of interest. The most prominent features are the *Capitol;* the *Central Pacific Railway Works;* the *Yolo Bridge;* some of the *Flouring Mills* and other manufactories. [Connections: southwestward to *San Francisco*, by rail or steamboat on the Sacramento; northward (by *Junction*) to *Marysville*, etc.]

Leave Sacramento by boat of the California Steam Navigation Company, down the Sacramento river to SAN FRANCISCO. Or,

Leave Sacramento by rail, by *Mokelumne Hill*, one of the oldest mining-places in California; by

STOCKTON, important town, commercial emporium of the southern mines, great wheat-centre, and lying at the head of navigation on the San Joaquin river; with fine view, northward, of *Monte Diablo*. [Connections: north to *Sacramento;* west to *Oakland* and *San José;* also by steamer to SAN

Francisco; also point of departure, by stage, for the Big Trees of Calaveras, the Yo-Semite Valley, Mariposa, etc.] Stockton to

Lathrop, junction of the Visalia Division of the Central Pacific railway. [Connection opening, by this route, to the Big Trees, the Yo-Semite, etc.] At beyond Lathrop is crossed the San Joaquin river, with views of the Contra Costa Mountains. By *Ellis*, *Niles* [connection southward, by rail, to the *Warm Springs of Alameda*]; by *Alameda*, on San Francisco Bay [connection with SAN FRANCISCO by rail and boat]; to

OAKLAND, large and pleasant town on San Francisco Bay, nearly opposite San Francisco, with fine shade of oaks, the *University of California*, and many educational institutions. From Oakland (*Oakland Point*), by railway ferry-boat to

SAN FRANCISCO.

Division D.

SAN FRANCISCO, WITH SHORT EXCURSIONS.

San Francisco, called the "Metropolis of the Pacific," as well as the "Golden City," lies on the western side of the Bay of the same name, with entrance to the Pacific Ocean through the "Golden Gate." It is immense in trade and wealth, with singularly-equable though sometimes-trying climate, and a dashing enterprize unparalleled elsewhere. *California*, *Montgomery*, *Clay*, and

Washington are among the principal streets; and *Market Street* divides them between north and south, as in Philadelphia. *Telegraph Hill*, at the northern side, gives a splendid view of the city and harbor, and of many of the distant mountains.

Among leading Public Buildings are the *U. S. Mint*, Commercial street (new one building, at Mission and Fifth streets); the *Custom House* (with Post office): *Merchants' Exchange*, California street; *New City Hall* (building,) Yerba Buena Park; *U. S. Marine Hospital*, Mission street; *Roman Cath. Orphan Asylum*, Market street; *St. Ignatius' College*, Market street; etc. Prominent Churches: *Grace Church* (Epis.); *St. Mary's* and *St. Patrick's Cathedrals* (Cath.); *Calvary Presbyterian; First Methodist; First Baptist; Jewish Synagogue Emanuel; Mariners' Church*, etc. Leading Theatres: the *California, Metropolitan, Maguire's Opera House*, Alhambra. Chinese Theatres: Dupont street and Jackson street. Leading Hotels: the *Grand, Occidental, Lick House* and *Cosmopolitan*.

Other Features of Interest will be found, *The Chinese*, whose head-quarters in the Western World are at San Francisco, and in whose "quarters" "Temples" and Theatres much experience may be gained; the *Great Sea Wall*, building along the water-front; the *Water Works*, etc.

Near Excursions will include those to *Lone Mountain Cemetery*, with fine outlook; to the *Cliff House*, (favorite drive or horse-cars); to the *Ocean House* and *Race Course*, near the latter; to the *Hunter's Point Dry Docks:* to the *Mission Dolores* (street car); to the *Presidio, Fort Point*, etc., (drive or street car); and many others, locally directed, for longer sojourners. There are also ferries to *Oakland, San Antonio, Alameda, Contra Costa, San Quentin* and *Saucelito*.

[Connections from San Francisco. (Local hotel-enquiry advisable, for particulars.) By rail to SACRAMENTO, *Marysville, Oroville, Shasta*, (Shasta Butte-Mountain) *Vreka* and other towns north; to *Stockton, San Jose, Visalia*, and other towns, and *New Almaden Mines*, south; eastward to SALT LAKE CITY, *Omaha* and the East, by route just traversed. (Railway being laid, farther north, to *Oregon City*, PORTLAND, *Vancouver*, etc.) By river-steamer to SACRAMENTO. By sea-steamer on the Pacific, to *Monterey, St. Luis, Santa Barbara, Acapulco*, and other towns on the Pacific, southward; with connection at PANAMA with the Panama Railway and steamers on the Atlantic from ASPINWALL to NEW YORK. Also by sea-steamer north to PORTLAND and other towns of Oregon. Also by sea-steamer to the *Sandwich Islands*, with connection thence to AUSTRALIA. Also by Pacific mail steamships to JAPAN and CHINA, with connections to BRITISH INDIA, the Peninsular and Oriental steamers and overland route to EGYPT, *Mediterranean* and EUROPE.]

Division E.

TO BIG TREES (BOTH GROUPS), AND YO-SEMITE VALLEY.

Leave San Francisco by rail to *Sacramento*. Then rail, by *Brighton, Florin, Elk Grove,* and *McConnell's*, to

GALT. At Galt take stage, by *Ione City, Jackson* and *Amador*, to

MOKELUMNE HILL. Mokelumne Hill, on horseback (no wheeled conveyance as yet practicable), to the

BIG TREES OF CALAVERAS, in the county of the same name, near the Stanislaus River. They are nearly 100 in number; 150 to 325 feet in height; diameter 10 to $30\frac{1}{2}$ feet; estimated age, 1,200 to 2,500 years. The largest in girth, the *Mother of the Forest*, is 61 feet in diameter at 6 feet from the ground; and the highest, the *Keystone State*, has a height of 325 feet. (For other names and particulars, depend on local guide, always in waiting. Hotel at the grove.)

Leave Big Tree Grove by stage to *Sonora* and

CHINESE CAMP. At latter place change to stage for *Big Oak Flat;* and thence on horseback to *Hardin's* and the *Lower Hotel* at the

YO-SEMITE VALLEY, on the Merced River, with scenery alleged to be more grand than any other on the globe, in many particulars. Special points of interest: *El Capitan*, gigantic separated rock;

the *Three Brothers*, also rocks; the *Bridal-Veil Fall*, 940, feet; the *Royal Arches*, rocks; the *Great Yosemite Fall*, in three leaps of 1,600, 434 and 600 feet; the *North* and *South Domes*, rocks; *Mirror Lake:* and the stupendous but frightful view of the whole Valley, from *Inspiration Point.* (Depend upon guide, necessary and always in readiness, for route and particulars.) Hotels at the Valley: *Lydig's, Black's* and *Hutchings'.* Proceed to *Clark's*, and thence make detour, a few miles, to the

BIG TREES OF MARIPOSA, with no less than 427 of the monsters, varying from 20 to 34 feet in diameter, and from 275 to 325 feet in height—many of them estimated to be 2,000 to 2,500 years old. Return to Clark's.

For return, horseback from Clark's to *White and Hatch's;* stage from White and Hatch's to *Mariposa* and *Modesta* (railway in progress); railway from Modesta to *Lathrop, Stockton,* and thence to SACRAMENTO or SAN FRANCISCO.

Shortest time necessary for this excursion, 6 days; advisable time, 8 to 10 days.

OFF-ROUTE AND MINOR PLACES.

[TOWNS AND OTHER PLACES NOT INDEXED OR MENTIONED IN ANY OF THE ROUTES, OR WITH OPTIONAL ROUTE HERE INDICATED.]

Adrian (Mich.) by rail from Detroit.
Afton (N. Y.) on Albany and Susquehanna road, from Albany or Binghamton.
Allentown (N. J.) from Trenton or Bordentown.
Amenia (N. Y.) Harlem railroad from New York.
Amherst (Mass.) from New London by New London Northern road.
Ansonia (Ct.) from Bridgeport by Naugatuck road.
Antietam [Battle Field] (Md.) from Harrisburg to Hagerstown; or from Harper's Ferry.
Appleton (Wis.) by rail from Milwaukie.
Ashley Falls (Mass.) from Bridgeport by Housatonic road.
Aspinwall (Isthmus, for *California*) from New York by Pacific Mail Steamers, 1st and 15th of every month.
Ashburnham (Mass.) from Fitchburg.
Atchison (Kansas,) by rail from Kansas city, (see this list.)
Aurora (N. Y.) on Cayuga Lake, (see this list.)
Avon (N. Y.) by rail from Rochester or Batavia.
Bainbridge (N. Y.) on Albany and Susquehanna road, from Albany or Binghamton.
Bath (Me.) by rail from Portland.
Bath (N. H.) from Wells River.
Bath (N. Y.) on Buffalo Division of Erie road, from Batavia or Corning.
Baton Rouge (La.) by steamer from New Orleans.
Beaufort (S. C.) from Charleston.
Belfast (Me.) by rail from Waterville, (see this list.)

Bennington (Vt.) by rail from Bellows Falls; or from Chatham Four Corners, (see this list.)
Benicia (Cal.) by boat from San Francisco.
Bethel (Me.) by rail from Portland or Gorham.
Bethel (Vt.) from White River Junction or Burlington.
Bethlehem (N. H.) from Littleton.
Beverly (Mass) by rail from Salem.
Bolton [and *Falls*] (Vt.) from Ridley's Station, (see this list.)
Booneville (Mo.) by rail from Jefferson City, (see this list); or from St. Louis, by boat.
Booneville (N. Y.) from Utica.
Bowdoin College (Me.) at Brunswick, (see this list.)
Bradford (Vt.) by rail from Wells River or White River Junction.
Brandon (Vt.) by rail from Rutland or Burlington.
Braintree (Mass.) from Boston by South Shore road.
Bridgewater (Mass.) from Boston by Old Colony road.
Bristol (Ct.) by rail from Waterbury, (see this list), or Providence.
Bristol (N. H.) from Concord by N. New Hampshire road.
Bristol (R. I.) by rail from Providence.
Brookfield, (Ct.) from Bridgeport by Housatonic road.
Brunswick (Me.) by rail from Portland.
Canaan (Ct.) from Bridgeport by Housatonic road.
Cairo (N. Y.) by stage from Catskill.
Camel's Hump [Mountain] (Vt.) by carriage from Ridley's Station, (see this list.)
Canton (Mass.) by rail from Providence or Boston.
Carlisle (Pa) by rail from Harrisburg.
Caseyville (Ill.) by rail from St. Louis.
Castine (Me.) by boat from Belfast, (see this list.)
Carbondale (Ill.) by rail from Cairo.
Centralia (Ill.) by rail from Cairo or Chicago.
Chateaugay Woods (N. Y.) from Rouse's Point, or from Plattsburg.

Chatham (N. J) by Morris and Essex road from New York.
Chatham (N. Y.) Harlem railroad from New York, or Boston and Albany road from Boston.
Charlemont (Mass.) from North Adams.
Charleston (S. C.) from New York by steamers twice a week or oftener.
Cheat River (W. Va.) by rail from Wheeling; or from Harper's Ferry.
Cheshire (Ct.) from New Haven by Northampton road.
Chester (Vt.) by rail from Bellows Falls.
Chicopee (Mass.) by rail from Springfield.
Chilicothe (O.) by rail from Cincinnati.
Circleville (O.) by rail from Cincinnati or Zanesville.
Clarendon [*Springs*] (Vt.) by stage from Rutland.
Clarksburg (W. Va.) by rail from Harper's Ferry; or from Wheeling by Grafton.
Clifton Springs (N. Y.) on Auburn Branch of New York Central road, from Syracuse or Rochester.
Coatesville (Pa.) from Philadelphia by Pennsylvania Central road.
Collinsville (Ct.) from New Haven by Northampton road.
Cooperstown (N. Y.) by Susquehanna road from Albany.
Crawfordsville (Ind) by rail from Indianapolis.
Crooked Lake (N. Y.) from Penn Yan, (see this list.)
Croton Falls (N. Y.) Harlem railroad from New York.
Dalles of St. Louis River (Minn.) by rail from St. Paul or Duluth.
Danbury (Ct.) from Norwalk.
Danielsonville (Ct.) by rail from New London or Worcester.
Deal (N. J.) from Long Branch.
Deerfield [and *South*] (Mass) by rail from Northampton, (see this list.)
Delaware (O.) by rail from Columbus.
Derby (Ct.) from Bridgeport by Naugatuck road.
Dexter (Me.) by rail from Bangor.

Dover Plains (N. Y.) Harlem railroad from New York.

Downington (Pa.) from Philadelphia by Pennsylvania Central road.

Easthampton (Mass.) from New Haven by Williamsburg road.

Eastport (Me.) by steamer from Boston and from St. John's, N. B.

Eatontown (N. J.) by New Jersey Southern road from New York; or from Long Branch.

Effingham (Ind.) by rail from Terre Haute or St. Louis.

Englewood (N. J) from New York by Northern New Jersey road.

Essex Junction (Vt.) from Burlington.

Falls Village (Ct.) from Bridgeport by Housatonic road.

Farmingdale (N. J.) by New Jersey Southern road (boat and rail) from New York.

Fitchburg (Mass.) by rail from Boston.

Fitzwilliam (N. H.) by rail from Fitchburg, Mass.

Flint (Mich.) by rail from Detroit.

Florence (Mass.) from New Haven by Williamsburg road.

Flushing (L I.) from New York by 34th street ferry and Flushing railroad.

Fond du Lac (Wis.) by rail from Milwaukie or Duluth.

Fort William (Canada), by boat from Duluth.

Foxborough (Mass.) by rail from Providence or Boston.

Franklin (Ind.) by rail from Indianapolis.

Franklin (N. H.) from Concord by Northern New Hampshire road.

Freehold (N. J.) from New York by Camden and Amboy or New Jersey roads, by Jamesburg; or from Long Branch.

Galveston (Texas), by steamer from New Orleans.

Gardiner (Me.) by rail from Portland.

Gettysburg (Pa.) by rail from Harrisburg, by York.

Glassboro (N. J.) by rail from Camden.

Gloucester (Mass.) by rail from Salem.

Grafton (N. H) from Concord by N. New Hampshire road.
Great Barrington (Mass.) from Bridgeport by Housatonic road.
Greensburg (Pa.) from the Pennsylvania Central road at Blairsville.
Greenfield (Mass.) from New Haven, by Northampton.
Greenport (Long Island), from New York by Long Island road.
Greenwich (Ct.) from New York by New Haven road.
Hackensack (N. J.) by Erie road from New York.
Hadley (Mass.) from Northampton, (see this list.)
Hagerstown (Md.) by rail from Baltimore, or Harrisburg by Chambersburg.
Hamilton (O.) by rail from Cincinnati.
Hammondsport (N. Y.) on Crooked Lake, (see this list.)
Hannibal (Mo.) by rail from Springfield, Ill.; or by river from St. Louis.
Hanover (N. H.) from White River Junction.
Havana (Cuba) from New York by Atlantic Mail Steamers, every Thursday; and by Vera Cruz Steamers, every 10 days.
Haydenville, (Mass.) from New Haven by Williamsburg road.
Highgate Springs (Vt.) from Rouse's Point.
Hingham (Mass.) from Boston by South Shore road.
Hinsdale (Vt.) opposite Brattleboro.
Holmdel (N. J.) by steamboat from New York to Keyport (see this list,) thence by stage.
Holyoke (Mass.) by rail from Springfield.
Honesdale (Pa.) by rail from Lackawaxen, on Erie Road.
Hoosac Tunnel (Mass.) from North Adams.
Housatonic (Mass.) from Bridgeport by Housatonic road.
Houston (Texas) by rail from Galveston (see this list.)
Howe's and *Ball's Caves* (N. Y.) from Schoharie (see this list.)
Ipswich (Mass) by rail from Salem.
Isle Royal (Mich.) by boat from Duluth

Ithaca (N. Y.) on Cayuga Lake (see this list.)
Jacksonville (Florida) from Savannah.
Jacksonville (Ill.) by rail from Springfield.
Jamaica (Long Island) from New York by Long Island road.
Jefferson City (Mo.) by rail from St. Louis.
Jonesboro (Ill.) by rail from Cairo.
Kane (Pa.) from Ridgeway or Irvineton, on Philadelphia and Erie road.
Kansas City, (Mo.) by rail from St. Louis, or from Omaha.
Kearsarge Mountain (N. H.) from Concord by Northern New Hampshire road.
Keene (N. H.) by rail from Fitchburg or Bellows Falls.
Kenoska (Wis.) by rail from Chicago.
Kent (Ct.) from Bridgeport by Housatonic road.
Keyport (N. J.) by steamboat from New York.
Killington Peak (Vt.) from Rutland.
Knightstown (Ind.) by rail from Dayton (O.) or Indianapolis.
Lafayette (Ind.) by rail from Logansport.
Lake Dunmore (Vt.) by stage from Brandon (see this list.)
Lake Luzerne (N. Y.) from Saratoga.
Lake Pleasant (N. Y.) from Amsterdam, New York Central road.
Lake Temisconata (Canada) from Riviere du Loup by Grand Portage road.
Lake Umbagog (Me.) from Gorham, N. H.
Lambertville (N. J.) from Trenton by Belvidere Delaware road.
Lancaster (O.) by rail from Columbus or Zanesville.
Lawrence, (Kansas) from Kansas City (see this list).
Lawrenceburg (Ind.) by rail from Cincinnati.
Leavenworth (Kansas), by rail from Kansas City (see this list.)
Lebanon (N. H.) from White River Junction.
Lee (Mass.) from Bridgeport by Housatonic road.
Lehigh Water-Gap and *Lehighton* (Pa.) from Easton by Lehigh Valley road.

Lenox (Mass.) from Bridgeport by Housatonic road; or from Albany by Boston and Albany road to Pittsfield.

Lexington (Mo.) by rail from Sedalia and Jefferson City (see this list).

Litchfield, (Ct.) from Bridgeport by Naugatuck road.

Little Rock (Ark.) by rail from Memphis, Tenn.

Logan (O.) by rail from Columbus.

London (O) by rail from Springfield.

Long Branch (N. J.) route opening, and to all other places in near connection, by "All Rail Route," from Rahway on the New Jersey road.

Ludlow (Vt.) by rail from Bellows Falls.

Mackinaw (Mich.) from Detroit, by boat.

Manches'er (Ct.) by rail from Hartford.

Manchester (N. J.) by New Jersey Southern road from New York.

Manchester (Vt.) by rail from Rutland.

Mansfield (Ct.) by rail from Hartford or Providence.

Marblehead (Mass.) by rail from Salem.

Marietta (O) by rail from Wheeling.

Martha's Vineyard (Mass.) by steamer from New Bedford.

Martinsburg (W. Va.) by rail from Harper's Ferry.

Martinsville (Ind.) by rail from Indianapolis.

Massena Springs (Canada), from Louisville, on the St. Lawrence, near Prescott.

Matawan (N. J.) by steamboat from New York to Keyport (see this list), thence by stage.

Mauch Chunk (Pa.) from Easton by Lehigh Valley road.

Maysville (Ky.) by rail, or the Ohio river from Cincinnati.

Meadville (Pa.) from Corry, Oil-Regions.

Medford (Mass.) from Boston by Lowell road.

Middleboro (Mass.) from Boston by Old Colony road.

Middleburg (Vt.) by rail from Burlington.

Middletown (N. J.) by New Jersey Southern road (boat and rail) from New York.

Middletown (Pa.) from Harrisburg.
Milford (Ct.) from New York by New Haven road.
Milford (O.) by rail from Columbus.
Minnesota Lakes (Minn.) from St. Paul.
Missisquoi Springs (Vt.) from St. Albans, by stage.
Mitchell (Ind.) by rail from Louisville.
Mound City [and *Mounds*] (Ill.) by rail from Cairo.
Monroe (Mich.) by rail from Detroit.
Montpelier (Vt.] from Burlington.
Mount Desert [and *Rock* and *Island*] (Me.) by steamer from Boston, Portland or Bangor.
Mount Diablo (Cal.) from San Francisco, by San Francisco and Oregon Railway, and connections.
Mount Holly (N. J.) by rail from Camden or Burlington.
Mount Holyoke (Mass.) from Northampton, (see this list.)
Mount Katahdin (Me.) by stage from Bangor; or partially by rail from same place.
Mount Mansfield (Vt.) by carriage from Waterbury, (see this list.)
Mount Tom (Mass.) from Northampton, (see this list.)
Mount Vernon (N. Y.) from New York by New Haven road.
Mount Vernon (O) by rail from Newark.
Nantucket (Mass.) by steamer from New Bedford.
Narragansett Pier (R. I.) from Kingston, on Stonington and Providence road.
Nassau (New Providence) from New York, by Atlantic Mail steamers, irregularly.
New Britain (Ct.) by rail from Waterbury (see this list) or Providence.
Newburg (O.) by rail from Cleveland.
Newburg (Vt.) by rail from Wells River.
New Egypt (N. J.) by rail from Hightstown, Mt. Holly or Burlington.
New Milford (Ct.) from Bridgeport by Housatonic road.

New Monmouth (N. J.) by New Jersey Southern road (boat and rail) from New York.
New Orleans (La.) from New York by steamers, every Saturday or oftener.
New Philadelphia (O.) by rail from Pittsburg.
New Rochelle (N. Y.) from New York by New Haven road.
Newtown, (L. I.) from New York by 34th street ferry and Flushing railroad.
Norfolk (Va.) from New York by steamer, every Saturday or oftener.
Normal (Ill.) by rail from Chicago.
North Adams (Mass.) from Bridgeport by Housatonic road, or from Boston, Albany or Troy by Troy and Boston road.
North Derby (Vt.) from Lenoxville, Canada, by Massiwippi road.
Northampton, (Mass.) by rail from New Haven.
Northfield (Mass.) from New London by New London Northern road.
Northfield (Vt.) from Burlington.
Northumberland (Pa.) from Harrisburg by Northern Central road.
Norwich (Vt.) by rail from White River Junction.
Oceanport (N. J.) by New Jersey Southern road from New York; or from Long Branch.
Oshkosh (Wis.) by rail from Milwaukie.
Otsego Lake (N. Y.) by Susquehanna road from Albany.
Otter Creek Falls (Vt.) from Vergennes (see this list.)
Oxford (Me.) by rail from Portland.
Oxford (O) by rail from Cincinnati.
Owasco Lake (N. Y.) from Auburn (see this list.)
Parkesburg (Pa.) from Philadelphia by Penn. Central road.
Parkersburg, (W. Va.) by rail from Wheeling, or from Columbus, O.
Passumpsic (Vt.) by rail from Wells River.
Phœnixville (Pa.) from Philadelphia by Reading road.

Patchogue (Long Island) from New York by South Side road.
Pemberton (N. J.) by rail from Hightstown, or from Camden, Burlington, or Long Branch.
Penn Yan (N. Y.) by Northern Central road, from Elmira.
Perth Amboy, (N. J.) from Rahway, by rail.
Peru (Ind.) by rail from Logansport.
Pittsfield (Mass.) from Bridgeport by Housatonic road, or from Albany or Boston by Boston and Albany road.
Piedmont (W. Va.) by Balt. and Ohio road from Harper's Ferry.
Pittsford (Vt.) by rail from Rutland.
Piqua (O.) by rail from Columbus.
Plainville (Ct.) from New Haven by Northampton road.
Pontiac (Ill.) by rail from Chicago.
Pontiac (Mich.) by rail from Detroit.
Port Huron (Mich.) by rail from Detroit.
Port Kent (N. Y.) by boat on Lake Champlain, going to or from Burlington, Vt.
Pottsville (Pa.) from Reading.
Poultney (Vt.) by rail from Rutland.
Putney (Vt) by rail from Bellows Falls.
Quincy (Mass.) from Boston by Old Colony road.
Racquette Regions (N. Y.) in connection with Adirondacks from Crown Point.
Ravenna (O.) by rail from Cleveland.
Readville (Mass.) by rail from Providence or Boston.
Red Bank (N. J.) by New Jersey Southern road (boat and rail) from New York.
Riceville (N. J.) by New Jersey Southern road from New York.
Richfield Springs (N. Y.) from Sharon Springs; or from Utica or Binghamton by the Utica, Chenango and Susquehanna road.
Richmond (Ind.) by rail from Xenia or Indianapolis.
Ridley's Station (Vt.) from Essex Junction, (see this list.)

Richmond (Va.) from New York by steamer every Saturday or oftener.
Rio Janiero (Brazil) from New York by United States and Brazil steamers, 23rd of every month.
Rockville (Ct.) by rail from Hartford.
Rockville (Ind.) by rail from Terre Haute.
Royalton [and *South*] (Vt.) from White River Junction.
Sackett's Harbor (N. Y.) by rail from Rome, on New York Central road.
Saginaw (Mich.) [and *East*] by rail from Detroit.
Salem (Ind) by rail from Louisville.
Salem (N. J) by rail from Camden.
Salem (N. Y.) by Troy and Boston road, from Troy.
Salisbury (Ct.) from Bridgeport by Housatonic road.
Salisbury [*East* and *Beach*] (Mass.) by rail from Salem.
San Francisco (Cal.) from New York by Pacific Mail steamers, 1st and 15th of every month.
San Rafael and *San Quentin* (Cal.) by boat from San Francisco, and horse.
Sault Ste. Marie (Mich.) [Rapids and Canal], by boat from Detroit.
Savannah (Ga.) from New York by steamer several times a week. [Connection for *Florida cities*.]
Schoharie (N. Y.) on Albany and Susquehanna road, from Albany or Binghamton.
Schuylkill Haven (Pa.) from Reading.
Scotch Plains (N. J.) by New Jersey Central road from New York.
Seabrook (N. H.) from Boston or Portsmouth.
Sedalia (Mo.) by rail from St. Louis.
Seneca Lake (N. Y.) on Auburn Branch of New York Central road, from Syracuse or Rochester.
Seymour (Ct.) from Bridgeport by Naugatuck road.
Shark River (N. J.) by New Jersey Southern road from New York.

Sharon (Vt.) from White River Junction.
Sheboygan (Wis.) by rail from Milwaukie.
Shelburne N. H. from Gorham.
Shelburne Falls (Mass.) from North Adams.
Shelbyville (Ind.) by rail from Indianapolis.
Sheffield (Mass.) from Bridgeport by Housatonic road.
Sheffield (Pa.) from Ridgeway or Irvineton, on Philadelphia and Erie road.
Shrewsbury (N. J.) by New Jersey Southern road from New York.
Skeneateles (N. Y.) on Auburn Branch of New York Central road, from Syracuse or Rochester.
Sorel (Canada), by steamer from Quebec.
South Hadley (Mass.) by rail from Springfield.
South Paris (Me.) by rail from Portland.
Squan [*Beach*] (N. J.) by New Jersey Southern road from New York to *Shark River*, thence stage.
Squankum (N. J.) by New Jersey Southern road from New York.
Stafford (Ct) by rail from Hartford or Providence.
Stanstead (Canada), from Newport, Lake Memphremagog, or from Lennoxville.
St. Augustine (Florida), from Savannah, Ga.
St. Charles (Mo.) by rail from St. Louis.
Sterling (Ct.) by rail from Hartford.
Stillwater (Minn.) from St. Paul.
St. John Falls (Canada), from Riviere du Loup or from Cacouna.
St. Johnsburg (Vt.) by rail from Wells River.
Stockbridge [and *West*] (Mass.) from Bridgeport by Housatonic road.
Stratford (Ct.) from New York by New Haven road.
St. Thomas (W. I) from New York by Brazil steamers, 23d of every month.
Sunbury (Pa.) from Harrisburg by Northern Central road.

Superior City (Wis.) by boat from Detroit or Chicago.
Sutherland Falls (Vt.) by rail from Rutland.
Terryville (Ct.) by rail from Waterbury (see this list) or Providence.
Thetford (Vt.) by rail from White River Junction.
Thompson (Ct.) by rail from New London or Worcester.
Thunder Bay (Lake Superior) by boat from Duluth.
Tiffin (O.) by rail from Sandusky.
Tolland (Ct.) by rail from Hartford or Providence.
Tom's River (N. J.) by New Jersey Southern road from New York.
Topeka (Kansas) from Kansas city (see this list.).
Troy (O.) by rail from Dayton.
Urbana (O.) by rail from Columbus.
Valley Forge (Pa.) from Philadelphia, or from Reading.
Vandalia (Ind.) by rail from St. Louis.
Van Deusenville (Mass.) from Bridgeport by Housatonic road.
Vera Cruz (Mexico), from New York by Mexican mail steamers, every 10 days.
Vergennes (Vt.) by rail from Burlington.
Vicksburg (Miss.) by river from New Orleans; or rail from Jackson.
Waltham (Mass.) from Boston by Fitchburg road.
Warren (O.) by rail from Cleveland.
Warren (R. I.) by rail from Providence.
Warren (Pa.) from Ridgeway or Irvineton, on Philadelphia and Erie road.
Waterbury (Ct.) from Bridgeport by Naugatuck road.
Watertown (Mass.) from Boston by Fitchburg road.
Waterville (Ct.) by rail from Waterbury, (see this list,) or Providence.
Waterville (Me.) by rail from Portland by Augusta.
Watkins Glen (N. Y.) from Elmira, Erie road.
Waukegan (Ill.) by rail from Chicago.

Waukesha (Wis.) by rail from Milwaukee.
Wellsville (O.) by Ohio river from Pittsburg.
West Burke (Vt) by rail from Wells River.
Westfield (Mass.) from New Haven by Williamsburg road.
Westminster (Vt.) by rail from Bellows Falls.
West Randolph (Vt.) from White River Junction or Burlington.
Weymouth (Mass.) from Boston by South Shore road.
White Plains (N. Y.) Harlem railroad from New York.
Wilkesbarre (Pa.) from Easton by Lehigh Valley.
Williamsburg (Mass.) from New Haven by W. road.
Williamstown (Mass.) from Albany, Troy or Boston, by Troy and Boston road.
Williston (Vt.) from Essex Junction.
Willoughby Lake (Vt.) from West Burke, (see this list.)
Wilmington (O.) by rail from Cincinnati.
Winstead (Ct.) from Bridgeport by Nangatuck road.
Woburn [East] (Mass.) from Boston by Lowell road.
Wolcotville (Ct.) from Bridgeport by Nangatuck road.
Woodbury (N. J.) by rail from Camden.
Woodstock (Vt.) from White River Junction.
Woodbridge (N. J.) from Rahway, by rail.
Woonsocket (R. I.) by rail from Providence or Worcester.
Wyandotte (Kansas.) opp. Kansas city (see this list).
Wyandotte (Mich.) from Detroit.
Wyoming (Minn.) by rail from St. Paul.
Wyoming Valley (Pa.) from Scranton, by the Lackawanna and Bloomsburg road.
Yankton (Dacotah), by the Missouri river from Sioux city.
Yellow Springs (O.) by rail from Xenia or Cincinnati.
York (Pa.) by rail from Harrisburg.

INDEX.

[PLACES AND ROUTES. FOR PLACES THEMSELVES, SEE FIGURES IN HEAVY TYPE; FOR PLACES NOT FOUND IN THIS INDEX, SEE "OFF-ROUTE AND MINOR PLACES," PRECEDING.]

A

Acapulco, Mex., 273.
Adirondack Mountains, 97, 100.
Adrian, Mich., **227**, 235.
Akron, Ohio, 210.
Alameda (and Warm Springs), Cal., 271, 273.
ALBANY, N. Y., 70.
Albion, Mich., 235.
Albion, N. Y., 77.
Alexandria, Va., **183**, 185, 195.
Alkali, Neb., 265.
Allegheny City, Pa., 200.
Allegheny Mountains, 199.
Allegheny Springs, Va., 195.
Allentown, Pa., 208.
Alliance, O., 210.
Altoona, Pa., 199.
Alton Bay, N. H., 125, 129.
Alton, Ill., **224**, 239, 261, 263.
Alton, N. H., 129.
Amador, Cal., 274.
Amherst, Mass., 139.
Ammonoosuc Falls, N. H., 134.
Anchorage, Ky., 217.
Andover, North, Mass, 124.
Annapolis Junction, Md., 171.
ANNAPOLIS, Md., **168**, 171.
Ann Arbor, Mich., 235.
Appomattox, Va., 195.
Aquia Creek, Va., 183, 185.
Argenta, Nev., 269.
Arlington House, Va., 182.
Arnprior, Can., 243.
Athabasca (Junction), Can., 128, 251.
Ashland, Iowa, 263.
Ashland, Ky., 217.
Aspen, Utah, 267.
Aspinwall, Cent. Amer., 273.
Athens, N. Y., 70.
Athens, O., 203.
Atlanta, Ga., 191.
Atlantic, Iowa, 264.
Atlantic City, N. J., 150, **160**.
Attleboro, Mass., 110.
Auburn, Cal., 270.
Auburn, N. Y., 76.
AUGUSTA, Ga., 191.
Augusta, Me., 127.
Au Sable River, N. Y., 100.
Austin, Iowa, 238, 261, 262.
Austin, Nev., 269.
Australia (to), 273.

B.

Ballston Spa, N. Y., 91.
BALTIMORE, Md., **164**, 198, 201—Fort McHenry, 165 — Monuments, 165 — Streets, 165—Public Buildings, 166—Churches 166—Lit. Inst., 166—Theatres, 167—Hotels, 167—Cemeteries, 167—Excursions, 168 — Longer Excursions, 168, 169, 170.
Bangor, Me., 127.
Barrytown, N. Y., 68.
Batavia, N. Y., 77.
Bathurst, N. B., 260.
Battle Creek, Mich., 235.
Battle Mountain, Nev., 269.
Bay City, Mich., 240.
Beauport, Can., 253.
Beaverton, Can., 242.
Becancour, Can., 128.
Bedford Springs, Pa., 199.
Bel-Air, Md., 168.
Belle Air, O., 202.
Belleville, Can., 243.
Bellows Falls, Vt., 140.
Belmont, Mo., 220.
Belmont, Nev., 269.

Beloit, Wis., 238.
Bergen Point, N. J.. 58, 149, 206.
Bergen Tunnel, N. J., 83.
Berlin, Can., 232, 242.
Bethlehem, N. H., 134.
Bethlehem, Pa., 298.
Beverley, N. J., 150.
Bidde.ord, Me., 125.
Big Oak Flat, Cal., 274.
Big Trees of Calaveras, Cal., 271, **274.**
Big Trees of Mariposa, Cal., 275.
Big Tunnel, Va , 195.
Binghamton, N. Y., 76, **87.**
Bird's Point, Mo., 221.
Birmingham, Pa., 200.
Bismarck, Mo., 221.
Bitter Creek, Wyo., 266.
Black River, Can., 128.
Bladensburg, Md., 171.
Blairsville, Pa., 200.
Bloody Pond, N. Y., 95.
Bloomington, Ill., **224,** 261, 262, 263.
Bloomsbury, N. J., 207.
Boca, Cal., 269.
Boise City, Idaho, 268, 269.
Bonsack's, Va.. 195.
Boone, Iowa, 262.
Boonton, N. J., 228.
Bordentown, N. J., 148, **150,** 160.
BOSTON, 107, 110, **115,** 251—Streets, 117—Public Grounds, 117 — Antiquities, 117 — Public Buildings, 118—Monuments, 119—Churches, 119—Libraries and Lit. Inst , 119—Theatres, 120—Hotels, 120—Excursions, 120—Harvard University, 120—Washington Head Quarters, 121—Mount Auburn Cemetery, 121—Longer Excursions, 121, 122.
Bothwell, Can., 232.
Bound Brook, N. J., 206.
Bowling Green, Ky., 219.
Bowmanville, Can , 242.
Bradford, Mass., 124.
Branch Intersection, Pa., 198.
Branchville, S. C., 191.
Brandywine Creek, Pa., 162.
Brattleboro, Vt , 140.
Breckenridge, Minn., 239.
Bridgeport, Ct., 103.
Bridgeton, N. J., 150, 159.
Brighton, Cal., 274.
Brighton, Mass., 107.

Bristol, Pa., 148.
Bristol, Tenn , 195.
Brockport, N. Y., 77
Brockville, Can., 245.
BROOKLYN connection with New York)—Streets, 47 — Churches, 50—Public Buildings, 51—Hotels 53—Theatres, 54—Churches for Service, 54—Public Grounds, 55, 57—Prospect Park, 57—Greenwood Cemetery, 57—Excursions 57, 58—Navy Yard, 58.
Brown's, Can., 246.
Brown University, 109,
Bryan, Wyo., 267.
Buckingham, Can., 246.
Bucyrus, O., 210.
Budd's Lake, N. J., 60, 228.
BUFFALO, N. Y., **89,** 232, 234, 240.
Buford, Wyo , 266.
Bull Run (Battle-field), Va., 183.
Bulstrode, Can., 128, 251.
Bureau, Ill., 263.
Burkeville (Junction), Va., 195.
Burlington, Iowa, 215, 238, 239, 261, 263, 264.
Burlington, N. J., **150,** 160.
Burlington, Vt., **100,** 101, 140.
Bush River Bridge, Md., 164

C.

Cacouna, Can., 255.
Cairo, Ill., 194, 205, 216, **220.**
Caldwell, N. Y., 95.
Caledonia, N, Y., 77.
Caledonia Springs, Can., 246.
California Junction, Iowa, 264, 265.
Callicoon, N. Y., 86.
Calvary Mountain, Can., 247.
Camden, N. J., 148, **150,** 159.
Camel's Hump Mountain, Vt., 100.
Canandaigua, N. Y., 77.
Canton, O., 210.
Cape Cod, Mass., 122.
Cape Elizabeth, Me., 125.
Cape May, N. J., 150. **160.**
Cape Rouge, Can., 253.
Cape Vincent, N. Y., 243.
Carbon, Wyo., 266.
Carbondale, Pa., 229.
Carillon, C n., 246.
Carleton Place, Can., 243, 245.
Carlin, Nev., 269.
Carlisle, Pa , 198.
Carson, Nev.. 269.

Carter, Utah, 267.
Cascade Bridge, N. Y., 87.
Casey, Iowa, 264.
Castle Rock, Utah, 267.
Castleton, N. Y., 70.
Castleton, Vt., 101.
Catawissa, Pa., 209.
Catskill Landing, N. Y., 68.
Catskill Mountain House, 69.
Cave City, Ky., 218.
Cayuga, N. Y., 77.
Cedar Rapids, Iowa, 215, 238, **261**, 263.
Central City, Col., 266.
Centre Harbor, N. H., 102, **130**, 138.
Chambersburg, Pa., 183, 198.
Champaign, Ill., 224.
Charles City, Iowa, 238.
Charleson, Mo., 221.
CHARLESTON, S. C., 184, **189**.
Charlotte, N. Y., 77.
Charlotte Town, P. E. I., 259.
Charlottesville, Va., **183**, 195.
Chatham, Can., 232.
Chattanooga, Tenn., 191, **196**.
Chaudiere Junction, Can., 128, 254, 255, 256.
Chaudiere Falls, Can., **244**, 255.
Chelsea, Mass., 122.
Chenoa, Ill., 224.
Cherry Valley, N. Y., 74.
Chester, N. J., 228.
Chester, Pa., 162.
Chester, Va., 186.
CHEYENNE, Wyo., 265.
CHICAGO, Ill., 201, 205, **212**, 220, 223, 225, 227, 234, 236, 238, 239.
Chicamauga, Tenn., 196.
China (to), 273.
Chinese Camp, Cal., 274.
CINCINNATI, O., 201, 202, **204**, 205, 210, 216, 221, 223, 227.
Claremont, Vt., 140.
Clearfield, Pa., 199.
Cleveland, O., 201, 202, 203, 205, 210, 216, **226**, 240.
Clinton, Iowa, 261.
Clyde, N. Y., 77.
Clyde, O., 227.
Coal Regions of Pennsylvania, 207, 208, 229.
Coatesville, Pa., 197.
Coburg, Can., 242.
Cohoes Falls, N. Y., 71.
Colborne, Can., 243.
Cold Spring, N. Y., 66.

Colfax, Cal., 270.
Collins' Bay, Can., 243.
Colmar, Iowa, 238.
Coloma, Cal., 270.
Columbia, Ind., 211.
Columbia, Pa., 198.
COLUMBIA, S. C., 188, **191**.
Columbia Springs (route to), 70.
Columbus, Ky., 220.
COLUMBUS, O., **202**, 210, 227.
Columbus, Neb., 265.
Communipaw, N. J., 206.
Como, Cal., 247.
Coney Island, N. Y., 58.
Concord, Mass., **123**, 136.
CONCORD, N. H., 124, 125, **137**.
Conemaugh Station, Pa., 199.
Contra Costa, Cal., 273.
Conway, N. H., 102, 129, **131**, 138.
Conway Valley, N. H., 130.
Cooperstown, N. Y., 74.
Corinne, Utah, 268.
Corning, N. Y., 77, **88**.
Cornwall Landing, N. Y., 66.
Corry, Pa., 89, 198, 226, 230.
Coshocton, O., 202, 227.
Council Bluffs, Iowa, **262**, 264.
Covington, Ky., 204, 217.
Coxsackie, N. Y., 70.
Cranberry, N. J., 150.
Crawford House, N. H., 133.
Cresson, Pa., 199.
Crestline, O., 210.
Creston, Wyo., 266.
Crisfield, Md., 163.
Croton River, N. Y., 63.
Culpepper, Va., 183.
Cumberland, Md., 170, 201.

D.

Danvers, No., Mass., 124.
Danville, Can., 128.
Danville Junction, Me., 127.
Darien, Ct., 103.
Davenport, Iowa, 239, **263**.
Dayton, O., 204, 210, 211, 227.
Delaware Water Gap, **60**, 160, 207, 229.
Dennison, O., 202.
DENVER, Col., 266.
Deposit, N. Y., 86.
DES MOINES, Iowa, 215, 262, **264**.
DETROIT, Mich., 210, 211, 216, 227, **233**, 240.
Devil's Gate, Utah, 267.

294 INDEX.

Dexter, Iowa, 264.
Dixon, Ill., 261.
Dobbs Ferry, N. Y., 63.
Dorsey, Md., 171.
Doncet's Landing, Can., 128.
Dover, Del., 163.
Dover, N. J., 228.
Dover, N. H., **125,** 129, 137.
Downieville Cal., 270.
Downington, Pa., 197.
Drakesville, N. J., 228.
Dresden Junction, O., 202.
Dubuque, Iowa, 239, 261.
Duluth, Minn., 238, **240.**
Dundas, Can., 232.
Dunkirk, N. Y., 89 226.
Dunleith, Iowa, 261.
Dunville, Can., 232.
Durham, N. H., 125.

E.

Easton, Pa., 160, **207.** 229.
East Penn. Junction, Pa., 208.
Eatontown, N. J., 59.
Ebensburg, Pa., 199.
Echo Canon (and City), Utah, 267.
Echo Lake, N. H., 141.
Egypt (to), 273.
Elgin, Ill., from Chicago, 212 to 216.
Elizabeth City, Va., 169.
Elizabeth, N. J., 59, **145,** 197. 206.
Elizabethport, N. J., 149, 206.
Elk Grove, Cal., 274.
Elkhart, Ind., 227.
Elkhorn, Neb., 265.
Elkhorn, Wis., 237.
Elko, Nev., 268.
Elkton, Md., 163.
Ellicott's Mills, Md., 170.
Ellis, Cal., 271.
Elmira, N. Y., **88,** 198.
Eminence, Ky., 217.
Emporium, Pa., 230.
Englewood, Ill., 262.
Erie Canal, 74.
Erie, Pa., 198, 201, **226,** 230, 240.
Essex Junction, Vt., 101.
Europe (to), by Pacific, 273.
Evansville, Ind., 219.
Exeter, N. H., 125.

F.

Fairfax Court-House, Va., 183.
Fall River, Mass., 113, **114.**

Falls of the Ammonoosuc, N. H., 134.
Falls of the Chaudiere, Can., **244,** 253.
Falls of Cohoes, N. Y., 71.
Falls of the Genesee, N. Y., 77.
Falls, Glen Ellis, N. H., 131.
Falls, Glenn's, N. Y., 95.
Falls, Kauterskill, N. Y., 69.
Falls of Minnehaha, Minn., 239.
Falls of Niagara, 77, **78,** 242.
Falls, Passaic, 59, 84.
Falls of the Potomac, D. C., 182.
Falls, Rideau, Can., 244, 245.
Falls of the Sawkill, N. Y., 86.
Falls of St. Anne, Can., 247.
Falls of St. Anthony, Minn., 239.
Falls of Trenton, N. Y., 75.
Falls of the Yo Semite, Cal., 275.
Farmington, N. H., 129.
Fishkill Landing, N. Y., 66.
Fitchburg, Mass., 140.
Flemington, N. J., 207.
Florence, S. C., 189.
Florin, Cal., 274.
Flume, The (and House), N. H., 142.
Fonda, N. Y., 74.
Forest, O., 210.
Fort Bridger, Utah, 267.
Fort Dodge, Iowa, 262, 264.
Fort Hamilton, N. Y., 58.
Fort Howard, Wis., 261.
Fort Massasauga, Can., 257.
Fort Niagara, N. Y., 257.
Fort Plain, N. Y., 74.
Fort Saunders, Wyo., 266.
Fort Schuyler, N. Y., 111.
Fort Sedgwick, Neb., 265.
Fort Snelling, Minn., 239.
Fort Washington, Va., 185.
Fort Wayne, Ind., 201, **211,** 224, 227, 235, 262.
Fortress Monroe, Va., 169.
Four Lakes, Wis., 238.
Foxboro, Mass., 110.
Framingham, Mass., 107.
Franconia Notch, N. H., 134.
Frankford, Pa., 148.
Franklinton, Md., 168.
Frederick, Md., 170.
Fredericksburg, Va., 185.
Fredericton, N. B., 259.
Freehold, N. J., 149.
Freeport, Ill., 237, 261, 263.
Freeport, Pa., 200.
Fremont, O., 227.

Fremont, Neb., 262, 265.
Frenchman's Bay, Can., 242.

G.

Galena, Ill., 239.
Galt, Cal., 274.
Galveston, Texas, 192, 194.
Gananoque, Can., 243.
Garrison's Landing, N. Y., 64.
Gatineau River, Can., 245.
Genesee Falls, N. Y., 77.
Geneva, N. Y., 77.
Genoa, Ill., 237.
Georgetown, Cal., 270.
Georgetown, D. C., 152.
Georgetown, Mass., 124.
Germantown, Pa., 159.
Gilmore, Neb., 265.
Girard, Pa., 226.
Glen Allen, Mo., 221.
Glencoe, Can., 232.
Glen Cove, L. I., 61.
Glen Ellis Falls, N. H., 131.
Glen House, N. H., 131.
Glenn's Falls, N. Y., 95.
Goderich, Can., 232, 240.
Golden City, Cal., 266.
Gold Hill, Nev., 269.
Gonic, N. H., 129.
Gordonsville, Va., 183, 186.
Gorham, N. H., 102, 127, 251.
Gosport Navy Yard, Va., 169.
Grafton, Mass., 107.
Grand Haven, Mich., 227, 235.
Grand Island, Neb., 265.
Grand Junction, Iowa, 262.
Grand Rapids, Mich., 235.
Granger, Utah, 267.
Granite Canon, Wyo., 266.
Grass Valley, Cal., 270.
Great American Desert, 268.
Great Bend, N. Y., 87, 229.
Great Falls, N. H., 125.
Great Nevada Sandy Desert, 269.
Great Salt Lake, Utah, 268.
Green Bay, Wis., 215, 261.
Greenbush, N. Y., 70.
Greenfield, Mass., 140.
Greenport, L. I., 61.
Green River, Wyo., 266.
Greensboro, N. C., 188.
Greensburg, Pa., 200.
Greenwich, East, R. I., 109.
Greenwood Lake, N. Y., 85.
Grenville, Can., 246.

Greycourt, N. Y., 85.
Grimsby, Can., 231.
Grinnell, Iowa, 263.
Groton, Ct., 108.
Groton Junction, Mass., 126, 140.
Grout's Corners, Mass., 140.
Guelph, Can., 232, 242.
Gunpowder River Bridge, Md., 164.

H.

Hackettstown, N. J., 60, 228.
Ha Ha Bay, Can., 256.
HALIFAX, N. S., 250, **259**.
Hamilton, Can., **231**, 241.
Hamilton, O., 210.
Hampton Junction, N. J., 207, 228.
Hampton, N. H., 125.
Hampton and Roads, Va., 169.
Hancock, N. Y., 86.
Hannibal, Mo., 269.
Hanover, Md., 171.
Hanover, Va., 163.
Harper's Ferry, W. Va., **170**, 171, 183, 201.
HARTFORD, Ct., 105.
Harrisburg, Can., 232.
HARRISBURG, Pa., 160, 163, **198**, 209.
Harvard University, 120.
Haverhill, Mass., 124.
Havana Cuba, 194.
Havre de Grace, Md., 162, **163**.
Hawkesbury, Can., 246.
Hazard, Wyo., 266.
Hazleton, Pa., 209.
Helena, Nev., 268.
Henderson, Ky., 219.
Herkimer, N. Y., 74.
Hermitage, the, Tenn., 219.
Hickford Junction, Va., 189.
Highlands of Navesink, 59.
Highlands of the Hudson, 63.
Hightstown, N. J., 150.
Hillsdale, Mich., 227.
Hillsdale, Wyo., 265.
Hoboken, N. J., 58.
Homewood, Pa., 210.
Honesdale, Pa., 229.
Hoosic Tunnel, Mass., 136.
Horicon, Wis., 237.
Hornellsville, N. Y., 89.
Hudson, Can., 247.
Hudson, N. Y., 70.
Hyannis, Mass., 122.
Hyde Park, N. Y., 67.

I.

Indiana, Pa., 200.
India (to), 273.
INDIANAPOLIS, Ind., **211**, 218, 223.
Ione City, Cal., 274.
Iowa City, Iowa, 263.
Island Pond, 127, 251, 253, 255.
Isle of Shoals, N. H., 125.
Iron Mountain, Mo., 221.
Ironton, Mo., 221.
Irvineton, Pa., 230.
Ithaca, N. Y., 88.

J.

Jackson, Cal., 274.
Jackson, Mich., 235.
JACKSON, Miss., 194.
Jacksonville, Ill., 224.
Jamaica, L. I., 61.
Jamesburg, N. J., 149.
Jamestown (ruins), Va., 160.
Janesville, Wis., 215, 237.
Japan (to), 273.
Jefferson City, Mo., 223.
Jeffersonville, Ind., 218
Jersey City, N. J., **144**, 197.
Johnsonville, Tenn., 220.
Johnstown, Pa., 199.
Joliet, Ill., 224, **262**.
Jonesville, Mich., 227.
Julesburg, Neb., 265.

K.

Kalamazoo, Mich., 227, 234, **235**.
Kauterskill Falls, 69.
Kearney (and Fort, Neb., 265.
Keene, N. H., 140.
Keeseville, N. Y., 100.
Kelton, Utah, 268.
Kenosha, Wis., 237.
Kennebunk, Me., 125.
Kensington, Pa., 148.
Keokuk, Iowa, 239, 261, 262, 264.
Kinderhook, N. Y., 70.
Kingston, Can., **243**, 257.
Kingston, N. Y., 67.
Kingsville, S. C., 191.
Kittery, Maine, 125.
Knoxville, Tenn., 188, **195**, 219.
Komoka, Can., 232.

L.

Lachine, Can., 247.
Lachine Rapids, Can., 250.
Lackawaxen, N. Y., 86.
L Crosse, Wis., 239.
Lafayette, Ind., 211.
Lafayette College, 208.
Lagrange, Ky., 217.
Lake Champlain, 99.
Lake Donner, Cal., 269.
Lake George, 95.
Lake Hopatcong, 60, 228.
Lake Mahopac, 60.
Lake Memphremagog, Can., 127, 138, 140.
Lake Pepin, Minn., 239.
Lake Ponchartrain, La., 194.
Lake Simcoe, Can., 242
Lake St. Charles, Can., 253, 254.
Lake St. Clair, 233.
Lake St. Louis, Can., 247.
Lake St. Peter, Can., 128.
Lake Superior, 215, 240.
Lake Superior Copper Regions, 240.
Lake Superior Iron Regions, 240.
Lake Tahoe, Cal., 269.
Lake Winnepesaukie, 125, **129**.
Lake of the Two Mountains, Can., 247.
Lambertville, N. J., 207.
Lamokin Junction, Pa., 162.
Lancaster, Pa., 160, **197**.
L'Ance a l'Eau, Can., 255.
Lanesborough, N. Y., 87.
LANSING, Mich., 235.
Lansingburgh, N. Y., 71.
Laporte, Ind., 227.
La Prairie, Can., 100.
Laramie, Wyo., 266.
La Salle, Ill., 263.
Lathrop, Cal., 271, 275.
Laurel, Md., 171.
Lawrence, Mass., 122, 124, 136.
Lawrenceville, Pa., 200.
Lawton, Mich., 236.
Lazaretto, The, Pa., 162.
Lebanon, Pa., 198, **209**.
Lebanon Springs, N. Y., 70.
Leesburg, Va., 183.
Lectonia, O., 210.
Lennoxville, Can., 135.
Le Roy, N. Y., 77.
Lewes, Del., 163.
Lewiston, N. Y., 242, 257.
Lewiston, Pa., 199.
Lexington (and Junction), Ky., 217.
Lima, O., 211, 227.
LINCOLN, Neb., 262.

INDEX. 297

L'Islet, Can., 254.
Little Falls, N. Y., 74.
Little Rock, Ark., 220.
Littleton, N. H., 102, **139**, 140.
Lock Haven, Pa., 199, **230**.
Lockport, N. Y., 77.
Logansport, Ind., 211, 224, 227.
London, Can., 232.
London, O., 203.
Long Branch, N. J., **59**, 149, 150, 161.
Long Sault Rapids, Can., 246.
Lookout Mountain, 196.
Lorette, Can., 253.
L'Original, Can., 246.
LOUISVILLE, Ky., 211, **217**, 221.
Loveland, O., 204.
Lowell, Mass., **121**, 124, 136.
Lundy's Lane, Can., 81.
Lyme, Ct., 108.
Lyme, E. & S., Ct., 108.
Lynn, Mass., 121, 122.
Lynchburg, Va., 195.
Lyons, N. Y., 77.

M.

Macon, Ga., 191.
Madison, N. J., 228.
MADISON, Wis., 215, **237**, 261.
Major's, Can., 246.
Malden, N. Y., 68.
Malden (So.) Mass., 122.
Mallory Town, Can., 243.
Mammoth Cave of Kentucky, 218.
Manassas Junction, Va., 183, 195.
Manchester, N. H., 124, **137**.
Mankato, Minn., 238.
Mansfield, Mass., 110.
Mansfield, O., 210.
Mantua Junction, Pa., 197.
Manunkachunk, N. J., 228, 229.
Marietta, O., 204.
Mariposa, Cal., 275.
Marshall, Iowa, 262.
Marshall, Mich., 235.
Martinsburg, Pa., 199.
Marquand, Mo., 221.
Marquette, Lake Superior, 240.
Marysville, Cal., 270, 273.
Mason City, Iowa, 262, 263.
Massillon, O., 210.
Mast Hope, N. Y., 86.
McGregor, Iowa, 238.
McKenzie, Tenn., 220.
McPherson (and Fort), Neb., 265.
Medford, Mass., 124.

Medina, N. Y., 77.
Memphis Junction, Ky., 219.
Memphis, Tenn., 194, 196, 219, 220.
Mendota, Ill., 263.
Mendota Junction, Minn., 238.
Meredith Village, N. H., 101, 138.
Meriden, Ct., 105.
Meridian, Miss., 196.
Merrimac River and Valley, 136.
Michigan City, Mich., 236.
Middlebury, Vt., 101.
Middletown, N. Y., 85.
Middletown, Pa., 198.
Milford, Va., 186.
Miliville, N. J., 150.
Milton, Pa., 229.
Milton, Wis., 237.
Milroy, Pa., 199.
Milwaukie, Wis., 215, **237**, 239, 261.
Mineral Point, Mo., 221.
Mingo Junction, O., 202.
Minneapolis, Minn., 238.
Minnehaha, Min., 238.
Mississippi, Mouths of, 194.
Missouri Valley Junction, Iowa, 262.
Mobile, Ala., **192**, 196.
Modesta, Cal., 275.
Mohawk Valley, N. Y., 73.
Mokelumne Hill, Cal., 270, 274.
Monmouth Junction, N. J., 149.
Monocacy (and Valley), Md., 170.
Monroeville, O., 227.
Monson, Mass., 139.
Monterey, Cal., 273.
MONTGOMERY, Ala., **191**, 196, 220.
Monticello, Va., 183.
Montmorenci, Falls of, Can., 253.
Montpelier, Vt., 140.
MONTREAL, Can., 138, 242, 243, 244, 245, **247**, 253, 255, 257.
Moosehead Lake, 127.
Moreau Station, N. Y., 95.
Morrison, Ill., 261.
Morristown, N. J., 60, **228**.
Morrow, O., 204.
Moscow, Iowa, 263.
Mound City, Ill., 220.
Mount Holly, N. J., 150.
Mount Hope, R. I., 113, 114.
Mt. Joy, Pa., 198.
Mt. Lafayette, N. H., 141, 143.
Mt. Mansfield, Vt., 100.
Mountain of Rigaud, The, Can., 246.

Mount Vernon. Va., **184,** 185.
Mount Washington, N. H., 131, 132.
Mount Webster, N. H., 133.
Mount Willard, N. H., 133.
Murray Bay, Can., 254.
Muscatine, Iowa, 263.
Mystic, Ct., 109.

N.

Nahant, Mass, 121.
Napanee, Can., 243.
Narrowsburg. N. Y., 86.
NASHVILLE. Tenn., 191. **219.**
Nashua, Mass., 122, 124
Natick, Mass., 107.
Natural Bridge, Va., 195.
Nauvoo, Ill., 239.
Nebraska City, Neb., 262.
Nevada, Cal, 270.
New Almaden Mines, Cal., 273.
NEWARK, N. J., 59, **144,** 197, 228.
Newark, O., 202.
New Baltimore, N. Y., 70.
New Bedford, Mass., 122.
New Brighton, S. I., 59.
NEW BRUNSWICK, N. J., **146,** 197.
New Buffalo, Mich., 236.
Newburgh, N. Y., 67.
Newburyport, Mass., 124.
Newcastle (and Junction), Del., 163.
Newcastle, Pa., 210.
New Durham, N. H., 129
NEW HAVEN, Ct., **104,** 107.
New Lisbon, O , 210.
NEW LONDON, Ct., **108,** 114, 136, 139.
Newmarket Junction, N. H., 125
Newmarket, N. H., 125.
NEW ORLEANS, La., **192,** 196, 223, 239.
Newport, Ky , 204.
NEWPORT, R. I., 111.
Newport, Vt., 138, 140.
Newton, N J., 228.
NEW YORK CITY, 42—Harbor, 42—Streets, 46—Museums, 47—Libraries, 47—Public Galleries, 47—Educational Institutions, 48—Monuments, 49—Antiquities, 49—Churches, 49—Public Buildings, 50—Commercial Buildings, 51—Private Dwellings, 52—Hotel Buildings (Hotels), 53—Theatres, 53—Churches for Service, 54—Public Grounds, 55, 56, 57—Central Park, 56—Excursions, 58 to 61 — High Bridge, 57 — Jerome Park, 57 — Longer Excursions, 59.
Niagara Falls, 77, **78,** 242.
Niagara (Village), N. Y., 257.
Niles, Cal., 271
Niles, Mich., 236.
Norfolk, Va., 163 168.
Norristown, Pa., 160.
North Adams, Mass., 140.
North Bend, Neb., 265.
North Platte, Neb., 265.
Norwalk, Ct., 103.
Norwich, Ct., **114,** 136, 139.

O.

Oak Hill, N. Y., 68.
Oakland, Cal, 270, **271,** 273.
Oakville, Can., 241.
Oberlin, O., 227.
Ogallala, Neb, 265.
OGDEN, Utah, 267.
Ogdensburgh, N. Y., 76, 244, 245, 258.
Oil City, Pa., 201, 230.
Oil Regions, Pa , 89, 201, 226, 230.
Old Man of the Mountain, N. H., 141.
Old Point Comfort, Va , 169.
Oleopolis, Pa., 230.
Omaha, Neb., 216, 223, 262, **264,** 275.
Ontonagon, Lake Superior, 240.
Orange Court House, Va., 183.
Orange, N. J., 228.
Oregon City, Oregon, 273.
Orient, L. I., 61.
Orleans, Island of, Can., 254.
Oroville, Cal., 273.
Orville, O., 210.
Osino, Nev., 268.
Oswego N. Y., 76.
Otsego Lake, N. Y., 74.
OTTAWA, Can , 243, **244,** 259, 258
Ottumwa, Iowa, 262, 264.
Overlook Mountain House, 67.
Owatona, Minn., 238
Owego, N. Y., 88.

P.

Paducah (and Junction), Ky., 220.
Painesville, Pa. 226.

Palatine Bridge, N. Y., 74.
Palisade, Nev., 269.
Palisades, the Hudson, 63.
Palmer, Mass., 106, 139.
Palmyra, N. Y., 77.
Panama Cent Amer., 273.
Paradise, Idaho, 269.
Paris, Can., 232.
Parma, Mich., 235.
Passaic Falls, 59, **84.**
Paterson, N. J., 58, **83.**
Pawtucket, R. I., 110.
Peaks of Otter Va., 195.
Pekin, Ill., 224.
Pemberton, N. J., 150.
Pensacola, Fla., 192.
Peoria, Ill., 211, 224, 263.
Pequop, Nev., 268.
Percy, Wyo., 266.
Perth Amboy, N. J., 149.
Peterboro, Can., 243.
Petersburg, Va., 189.
Petrolia, Can., 232.
PHILADELPHIA, Pa., **151**-Streets, 152—Public Buildings, 153—Antiquities, 154—Public Grounds, 154—Fairmount, 154—Churches, 156—Libraries, 156—Theatres, etc., 157—Hotels, 157—Excursions, 157—Navy Yard, 157—Arsenals, 158—Laurel Hill Cemetery, 158—The Wissahickon, 158—Germantown, 159—Red Bank and Fort Mifflin, 159—Penn's Rock, 159—Falls of the Schuylkill, 159—Longer Excursions, 160.
Philipsburg, N. J., 207.
Pictou, N. S., 260.
Pictured Rocks, Lake Superior, 240.
Piermont, N. Y., 63.
Pike's Peak, Col., 266.
Pilot Knob., Mo., 221.
Pine Bluffs, Neb., 265.
Pinkham Notch, N. H., 131.
Pithole (and Creek) Pa., 230.
PITTSBURG, Pa., 170, **200**, 202, 226, 262.
Pittsfield, Mass., 139.
Piston, Pa., 208, 229.
Placerville, Cal., 270.
Plainfield, Ct., 115
Plainfield N. J., 206.
Plattsburg, N. Y., 99.
Plum Creek, Neb., 265
Plymouth, Ind., 211.

Plymouth, Mass., 122.
Plymouth, N. H., 138.
Plymouth, Wis., 238.
Pointe-aux Anglais, Can., 247.
Point Levi, Can., 128, 251, 254.
Point of Rocks, Md., 170.
Point St. Charles, Can., 249, 250.
Pollard, Ala., 192.
Pond Creek, Ill., 263.
Portage City, Wis., 237.
Portage, N. Y., 89.
Port Deposit, Md., 162.
Port Hope, Can., 242.
Port Huron, Mich., 234.
Port Jervis, N. Y., 85.
PORTLAND, Me., 102, **126**, 250, 251, 253, 255.
PORTLAND, Oregon, 268, 273.
Port Sarnia, Can., 232.
Portsmouth, N. H., 125, 137.
Portsmouth, Va., (Naval Depot), 169.
Port Stanley, Can., 232.
Potomac Falls, D. C., 182.
Potosi, Mo., 221.
Poughkeepsie, N. Y., 67.
Prairie du Chien, Wis., 215, **238**, 239.
Prescott, Can., **243**, 245, 258.
Princeton, N. J., 147.
Profile House, N. H., 134, 139, 140, **141.**
Promontory Point, Utah, 268.
PROVIDENCE, R. I., **109**, 115.
Put-in-Bay Islands, (Lake Erie), 234.
Putnam, Ct., 115.

Q.

QUEBEC, Can., 128, 138, 250, **251**, 255, 256.
Queenston, Can., 81, 257.
Quincy, Ill., 224, 263

R.

Racine, Wis., 237.
Rahway, N. J., **146**, 197.
RALEIGH, N. C., 189.
Ramapo Gap and Valley, N. J., 85.
Ramsey, Minn., 238.
Reading, Mass., 124.
Reading, Pa., 197, 198, 208, **209**, 229.
Readville, Mass., 110.

Red Bank, N. J., 59
Red Wing, Minn., 239.
Reno, Nev., 269.
Renovo, Pa., 230.
Rhinebeck, N Y, 67.
Rice Lake, Can, 243.
Richmond, Can, 128, 251 253, 255.
Richmond, O., 204.
RICHMOND, Va., 169, 183 **186,** 205.
Rideau Falls, Can., 244, 245.
Ridgeway Junction, N. C., 189.
Riviere du Loup, Can., 128, 251, 253, **254,** 256
Rochester, N. H., 125, 129
Rochester, N. Y., 77
Rochester, Pa., 202, 210.
Rockaway, N. J., 228.
Rockaway, N. Y., 58.
Rockford, Ill. 237.
Rock Island, Ill., 225 239,261,**263.**
Rockland Lake, 63.
Rome, N. Y., 76.
Rondout, N. Y., 67.
Rouse's Point, Vt., **100,** 140, 244, 250.
ROUTES:
No. 1.—Northern. New York to West Point, Catskill, Albany, Troy, Utica, Trenton Falls Rochester and Niagara Falls—62 to 82.
No. 2.—Northern. New York to Paterson, Upper Delaware, Upper Susquehanna, Binghamton, Elmira, Buffalo and Niagara Falls (Erie Railway)—83 to 90
No. 3.—Northern. New York to Albany, Saratoga, Lake George, Adirondack Mountains, Lake Champlain, Vermont cities, and Montreal—91 to 103
No. 4.—Eastern. New York to New Haven, Hartford, Springfield, Providence, Newport, New London, Stonington and Boston (options)—103 to 123.
No. 5.—Eastern Boston to Portsmouth Portland, White Mountains. Quebec or Montreal—124 to 128.
No. 6.—Northern and Eastern. Boston to Lake Winnepesaukie, White Mountains, Portland and Canadian cities—129 to 135.
No. 7.—Northern and Eastern. New York to New London, Norwich, Worcester, New Hampshire cities, Lake Winnepesaukie and the White Mountains—136 to 143.
No. 8.—Near Western. New York to Newark, New Brunswick, Trenton and other New Jersey cities, and Philadelphia—144 to 161.
No 9.—Western and Southern. Philadelphia to Wilmington (Del.), Baltimore, Washington, and Richmond—162 to 188.
No. 10.—South Western. Richmond to Raleigh, Wilmington, (N. C.). Columbia (S. C), Charleston, Atlanta, Montgomery, Mobile and New Orleans—189 to 194.
No. 11.—South-Western. Richmond to Lynchburg Knoxville, Chattanooga, Mobile and New Orleans—195 to 196.
No. 12.—Western. New York or Philadelphia to Harrisburg, Pittsburg, Wheeling, Columbus and Cincinnati, (Penn. Cent. R. R.)—197 to 205.
No. 13.—Western. New York to Plainfield, Somerville, Easton, Allentown, Harrisburg. Pittsburg, Fort Wayne and Chicago, (Allentown Route)—206 to 216.
No. 14.—Western. Cincinnati to Louisville. Mammoth Cave of Ky. Nashville, Cairo, St Louis, and Chicago—217 to 225
No 15.—Northern and Western. Buffalo to Erie Cleveland, Toledo and other Ohio cities, Cincinnati or Chicago—226 to 227.
No 16.—Northern. New York or Philadelphia to Dover, Morristown, Delaware Water-Gap, Scranton (coal regions), William-port and Lock Haven, (lumber regions), Oil City, Titusville, &c., (oil regions.)—228 to 230
No. 17.—Canadian and Western. Niagara Falls to St. Catherine's, Hamilton, Paris, London, Canada Oil-Regions, Detroit, Ann Harbor Kalamazoo and Chicago—231 to 236.
No. 18.—North-Western. Chicago to Racine Milwaukie, Madison,

INDEX. 301

Prairie du Chien, St. Paul, and Falls of Minnehaha and St Anthony, (with optional return by the Mississippi or Lake Superior)—237 to 240.
No. 19.—Canadian. Niagara Falls to St. Catherine's, Hamilton, Toronto, Kingston, Prescott, Ottawa, Montreal, Quebec and the Saguenay River—241 to 256.
No. 20.—Canadian. Niagara Falls to Toronto, Kingston, Thousand Islands, Rapids of the St. Lawrence, Montreal and Quebec, by steamers; and to Halifax, St. John, Fredericton, Windsor, Sidney, Shediac, Charlotte-Town, Pictou and Bathurst — 257 to 260.
No. 21.—Far Western. Chicago to Council Bluffs, Omaha, Cheyenne, Ogden Salt Lake City, Sacramento, San Francisco, the Big Trees, Yosemite Valley, &c. —261 to 275.
Rupert, Pa., 229.
Rutland, Vt., **101**, 140.
Rye Beach, N. H., 125.

S.

SACRAMENTO, Cal., **270**, 273, 274, 275.
Saginaw, Mich., 234.
Saguenay River, Can., 255.
Salamanca, Pa., 89, 226, 230.
Salem, Mass., 122.
Salem, O., 210.
Salisbury, Md., 163.
Salmon Falls, N. H., 125.
Salmon-Trout Lake, Can., 243.
SALT LAKE CITY, Utah, 216, 223, **267**, 273.
San Antonio, Cal , 273.
Sandusky, O., 202, 205, 210, 227.
Sandwich Islands, 273.
Sandy Hook, 59.
SAN FRANCISCO, Cal., 216, 223, 270, **271**, 275. Streets, 271.—Public Buildings, 272.—Churches, 272.—Theatres, 272.—Hotels, 272 — Excursions, 272.—Lone Mountain Cemetery, 272.—Cliff House, 272.—Mission Dolores, 273.—Presidio and Fort Point, 273.—Longer Excursions, 273.
San Jose, Cal., 270, 273.

San Luis, Cal., 273.
San Quentin, Cal., 273
Santa Barbara, Cal., 273.
Santa Fe, New Mex., 266.
Saranac Lakes, N. Y., 100.
Saratoga Springs, 91, **92**, 140.
Sancelito, Cal., 273.
Sault St. Marie, 240.
Savage, Md., 171.
Savannah, Ga., 190.
Sawkill Falls, N. Y., 86.
Schenectady, N. Y., 74.
Schooley's Mountain, N.J., **60**, 228
Schroon Lake, N. Y., 96
Schuyler, Neb., 265.
Scranton, Pa., 207, 208, **229**.
Seconnet (Point), R. I., 112.
Sexton's Junction, Va , 186.
Sharon Springs (route to), 74.
Shasta (and Butte), Cal., 273.
Shawangunk Mt., N. Y., 85.
Shediac, N. B., 259.
Shelbyville, Ky., 217.
Sherbrooke, Can., 127.
Sherman, Wyo., 266.
Shohola, N. Y., 86.
Sidney, C. B., 259.
Sidney, Neb., 265.
Silver City, Idaho, 269.
Sing Sing, N. Y., 63.
Sioux City, Iowa, 262, 264.
Skowhegan, Me., 127.
Sloatsburg, N. Y., 85.
South Amboy, N. J , 149.
South Bend, Ind., 227, 235.
South Berwick Junction, Me., 125.
South Pass, Wyo., 267.
South Reading (Junction), Mass., 124.
South Trenton, N. Y., 75.
South Vernon, Vt., 140.
Somerville, Mass., 122.
Somerville, N. J., 207.
Sonora, Cal., 274.
Sparta, Ky., 217.
Spotswood, N. J., 149.
SPRINGFIELD, Ill., **224**, 227, 261, 263.
Springfield, Mass., 106.
Springfield, O., from Xenia, 203.
Stafford, Ct., 139.
St. Albans, Vt., **101**, 244.
Stamford, Ct., 103.
Stanhope, N. J., 228.
St. Anne's, Can., 247.
St. Anne Falls of, Can., 247, **254**.
Stanstead, Can., 127.

St. Anthony (and Falls of), Minn., 215, **239**.
Starruca Viaduct, N. Y., 87.
Staten Island, 59.
Staunton, Va., 186, 195.
St. Catharine's, Can., **231**, 241.
St. Clair River and Lake, 240.
St. Cloud, Minn., 239.
Steubenville, O., **202**, 210.
Stevenson, Ala , 219.
St. Gregoire, Can., 251.
St. Hyacinthe, St. Brune, St. Hilaire, St. Lambert, Can., 128, 250.
St. John, Can., 100.
St. John, N. B., 259.
St. Joseph, Mo., 223, 262, 264.
St. Lawrence Rapids, 258.
St. Lawrence River, 258.
ST. LOUIS, Mo., 194, 201, 205, 216, 220, **221**, 227, 239, 261, 263.
St. Mary's, Can., 232.
St. Mary's, Wyo , 266.
Stockton, Cal , **270**, 273, 275.
Stonington, Ct., **109**, 115.
Stony Point, N. Y., 63.
St. Paul Junction, Minn, 238
ST. PAUL, Minn., 215, **238**, 262.
St Peter, Minn., 238, 239.
Strasburg, Va., 183.
Stroudsburg, Pa., 229.
St. Thomas, Can., 254.
Sturgis, Mich., 227.
Stuyvesant, N. Y., 70.
Suffern's, N. J , 84.
Summit Station, Cal., 269.
Summit, Va., 185.
Sunbury, Pa., 199.
Susquehanna, N. Y., 87.
Swampscott, Mass., 122.
Sweetwater Mining Region, Wyo., 267.
Syracuse, N. Y., 76.

T.

Tadoussac, Can., 255.
Tallahassee, Fla. 190.
Tarrytown, N. Y., 63.
Terre Haute, Ind., 220
Thorold Can., 231, 241
Thousand Islands (and Lake), St Lawrence River, 258.
Three Rivers, Can., 128, 251.
Throg's Neck, N. Y., 111.
Thurso, Can., 246.
Ticonderoga, N. Y., 98.

Tideoute, Pa., 230.
Titusville, Pa., 230.
Tiverton, R. I., 114.
Toano, Nev., 268.
Toledo, O., 203, 205, 210, 211, 216, **227**, 235, 240.
Tolland, Ct., 139.
TOPEKA, Kas , 223, 264.
TORONTO Can., 232, **241**, 243, 250, 257.
Townsend, Md., 163.
Towsontown, Md., 168.
Trenton, Can., 243.
Trenton Falls, 75.
TRENTON, N. J., **147**, 150, 197.
TROY, N. Y., 71.
Truckee, Cal., 269.
Tuscaloosa, Ala., 196.
Tyrone, Pa., 199.

U.

Uintah, Utah, 267.
Union City, Tenn., 220.
Union College (Schenectady, N. Y), 74.
Unionville, N. H., 129.
University of Virginia, 183.
Urbana, O., from Xenia, 203.
UTICA, N. Y., 74.

V.

Valparaiso, Ill., 211.
Vancouver, Ore ;on, 273.
Venango, Pa, 201.
Victoria Bridge, Can., 249, 250.
Vincennes, Ind., 205, 220.
Vineland, N. J., 160.
Virginia City, Nev., 269.
Virginia, Nev., 268.
Visalia, Cal., 273.
Vreka, Cal., 273.

W.

Wadsworth, Nev., 269.
Walla Walla, Oregon, 268.
Wallingford, Ct., 105.
Walton, Ky., 217.
Wanatah, Ind., 211.
Warrenton (and Junction), Va., 183.
Warsaw, Ind., 211, 224.
Wasatch, Utah, 267.
WASHINGTON, D. C., **172**, 201-Location, 172, 173—The Capitol,

174—President's House, 178.
—Patent Office, 178—Departments, 178, 179—Smithsonian Institute, 180—Public Grounds, 180—Churches, 181—Theatres, 181—Hotels, 181—Excursions, 181—Soldiers' Home, 181—Navy Yard 181—Congressional Cemetery 181—Longer Excursions, 181, 182.
Washington Junction, D. C., 171.
Washington, Iowa, 263.
Washington (Junction), N. J., 228.
Washington, N. J. 149.
Washoe, Nev., 269.
Waterford, Ct., 108.
Waterloo, Iowa, 261.
Waterloo, Ind., 211.
Waterloo, N. J., 228.
Watertown, N. Y., 76, 243.
Watertown, Wis., 237.
Waukegan, Ill., 237.
Waverley, Tenn., 220.
Waynesburg, Pa , 197.
Weber Canon, (and Station and River). Utah, 267.
Webster, Mass., 115.
Weir's Landing, N. H., 137.
Weldon, N. C., 188.
Wells (and Humboldt W.) Nev., 268.
Wells, Me., 125.
Wells River, Vt., 101, **138**, 140.
Wenona, Mich., 234.
Westerley, R. I., 109.
West Island, R. I., 112.
West Liberty, Iowa, 263.
West Newton, Mass., 107.
West Point, N. Y., 61, 62, **65**.
West Point, Ga., 191.
West Scarboro, Me., 125.
Wethersfield, Ct., 105.
Wheeling, W. Va., **201**, 210, 227.
Whitehall, N. Y., **101**, 140
White House, Va., 188.
White Mountain House, N. H., 134.
White Mountains, 125, 127, 130 to 133, 137, 141 to 143.
White Oak Bottom, Md., 171.
White Pigeon, Mich., 227.
White Pine, Nev., 268.
White River Junction, Vt., 102, 138, **140**.
White Sulphur Springs, Va., **184**, 186, 195.
Wickford, R. I., 109.

Wilcox, Pa., 230.
Wilkesbarre, Pa., 229.
Willey House, N. H., 133.
William-and-Mary College, Va., from Richmond, 186.
William's Bridge, N. Y., 103.
Williamsburg, Va., from Richmond, 186.
Williamsport, Pa., 198, **229**.
Willimantic, Ct., 139.
Wilmington, Del., **163**, 197.
Wilmington Junction, Mass., 124.
Wilmington, N C., 188, **189**.
Wilton, Iowa, 263.
Winchester, Va., 183.
Windsor (and Locks), Ct., 105.
Windsor, Can., 233.
Windsor, N. S., 259.
Windsor, Vt., 140.
Winnemucca, Nev., 269.
Wolfboro, N. H., 129, 138.
WORCESTER, Mass., **106**, 115, 136.
Wyandotte, Kas , 264.
Wyoming, Nev., 269.

X.

Xenia, O., 203.

Y.

Yale College, 104.
Yarmouth Junction, N. H., 102, 127.
Yarmouth, Mass., 122.
Yonkers, N. Y., 63.
York, Pa., 198.
Yorktown, Va , from Richmond, 186.
Yo Semite Valley, Cal., 271, **274**.
Ypsilanti, Mich., 234.

Z.

Zanesville, O., 202, 210, 227.

DISTANCES, TIME AND FARES.

[FROM NEW YORK, DIRECT BY RAIL TO MOST IMPORTANT POINTS—IN ROUND NUMBERS AND LIABLE TO SLIGHT VARIATION.]

NEW YORK TO	DISTANCES.	TIME.	FARES.
Albany	150 miles	5 hours	$ 3 50
Atchinson, Kan	1,370 "	60 "	46 20
Baltimore	200 "	8½ "	6 50
Baton Rouge	1,940 "	5 days	55 00
Boston	240 "	9¼ hours	6 00
Buffalo	450 "	15 "	9 50
Burlington, Iowa	1,125 "	50 "	32 30
Cairo, Ills	1,145 "	55 "	36 00
Charleston, S. C	800 "	48 "	26 00
Chattanooga	1,230 "	60 "	34 00
Chicago	915 "	38 "	25 00
Cincinnati	760 "	30 "	22 50
Cleveland, O	595 "	23 "	15 00
Corry (Oil Regions), Pa	520 "	16 "	11 70
Denver, Col	2,040 "	4½ days	95 70
Detroit	705 "	38 hours	16 50
Erie, Pa	550 "	17 "	12 25
Indianapolis	820 "	35 "	25 00
Mobile	1,600 "	86 "	50 50
Montreal	650 "	18 "	12 50
New Orleans	1,650 "	80 "	55 00
Niagara Falls	450 "	15 "	9 50
Omaha	1,413 "	3 days	44 50
Philadelphia	90 "	3½ hours	3 25
Pittsburg	445 "	16 "	13 00
Portland, Me	400 "	14½ "	9 50
Quebec	825 "	25 "	16 50
Quincy, Ills	1,147 "	51½ "	34 25
Richmond, Va	360 "	22 "	15 00
Salt Lake City	2,464 "	5 days	124 50
San Francisco	3,200 "	7½ "	140 00
Saratoga	225 "	7½ hours	4 50
Savannah	1,000 "	2½ days	34 50
St. Louis	1,084 "	50 hours	36 00
St. Paul	1,388 "	3 days	43 00
Washington	226 "	10 hours	7 00
White Mountains	500 "	20 "	8 00

STEAMSHIP LINES AND BANKING-HOUSES.

In arranging for visits to America, the European traveler needs especially to look after two points of no minor consequence. *First*, the character of the vessel in which he designs to make the voyage; and, *Second*, the standing and reliability of the banking-house from which he takes the exchange necessary for his disbursements while abroad. Upon both these points, it is the intention of the "Short-Trip Guide" to offer directions of importance, which cannot be ignored without disadvantage. A brief resumé follows, of steamship lines offering reliable and first-class conveyance to America; as well as of those banking-houses affording exchange upon such terms as will be found desirable to the traveler, and at the same time of such thorough reliability that they can be depended upon under all circumstances. Of steamship lines, let it be understood that *no vessel will have place here, in the way of commendatory announcement, not held by the compiler to be worthy of entrusting to it his own life;* that no banking-house will be recommended, without personal knowledge of its unimpeachable standing; and that in this department, as in all others of the volume, no dead enterprizes are allowed to remain announced from year to year, but all is current as well as reliable.

STEAMSHIPS TO AMERICA.

LINES FROM LIVERPOOL TO NEW YORK.

Cunard Line. (British and North American Royal Mail Steamship Company.)

The array of ships offered by this leading line, for the current year, is even more extensive than in past years, during which it has won the confidence of the world. The favorite "Scotia" (known as the "women and children's ship," from her steadiness); the speedy "Russia;" the "Cuba," "Java," "China," "Abyssinia," "Algeria," "Larthia," "Calabria," and other well known ships, are to be supplemented by the "Scythia" and "Bothnia," larger and finer than any of the others, and of great power and speed. Rates of passage varied to suit all purses. Sailings from Liverpool, Saturdays, Tuesdays and Thursdays; and from Queenstown (Ireland), the days following.

Inman Line. (Liverpool, New York and Philadelphia Steamship Company.)

New and fine ships are also the order of the day with the Inman line, which has so rapidly made popularity and holds it so well. The favorite ships of the line, of the last two or three years, the "City of Brussels," "Paris," "Brooklyn," "London," and others, have already been supplemented by the gigantic and powerful "City of Montreal;" and she is to be soon followed by the Commodore's ship, the "City of Chester," among the largest afloat and expected to be among the speediest—the "Richmond" and other fine vessels—making the fleet equally extensive and perfect. Sailings from Liverpool, on Thursdays and Tuesdays; and from Queenstown the days following.

Williams and Guion Line. (Liverpool and Great Western Steamship Company.)

A single vessel, the "Colorado," drops off from the rapidly-increasing number of vessels of this comparatively new but popular line; but the loss becomes a gain, in the immediate supply of the still larger and finer "Montana," very soon to be followed by the "Dacotah," and the latter by others now building, of the same noble class as the two last mentioned. No finer vessels, meanwhile, can be found, than the "Wyoming," the "Wisconsin" "Idaho," and other ships already on the route and supplying accommodation winning exceptional applause. Sailings from Liverpool, every Wednesday; and from Queenstown the following day.

National Line. (National Steam Navigation Co.)

In its earlier days the National line, while supplying safe ocean transit at lower rates than could be afforded by any other, had the reputation of using more time in the transit, owing to the slighter power of ships, than always pleased the hurried. But while gathering one of the largest fleets in existence, this line is also reversing past reputation by supplying some of the most powerful as well as largest ships in the world—the "Spain," "Egypt," "Italy" and other new vessels, being actual "flyers" as well as splendid monsters; and the "Queen," "England" "Denmark" and half a score of other vessels, following closely. Sailings from Liverpool, every Wednesday, and Queenstown the following day; with extra steamers.

White Star Line. (Ismay, Imrie & Co's; Oceanic Steamship Company.)

No other line ever built reputation so fast in a single year, as the White Star, with their immense ships and "all the modern improvements." The "Oceanic," pioneer of the line, made much reputation by carrying over so successfully the American Knight Templars, last summer; and she, and her successors, the "Atlantic," "Baltic" and "Republic," have all won applause for perfect fittings, comfort and quick passages. Upon these four still follow the "Celtic," "Adriatic," and several other and still larger vessels on the same novel and luxurious plan, to make the line complete and fully supplied. Sailings from Liverpool, every Thursday; and from Queenstown the following day.

LINE FROM GLASGOW AND LONDONDERRY TO NEW YORK.

Anchor Line. (*Henderson Brothers, New York: Handyside & Henderson, Glasgow.*)

This line has a rapidly-increasing fleet of fine steamers, of which the "Australia," "India," "Anglia," "California," "Europa," etc., may be named as among the favorites, with the "Bolivia" and "Utopia" to take their places during the season, and the Clyde ship-yards always busy in increasing the list. Round trips, by the way, embracing the Mediterranean ports as well as the Atlantic passage, have been arranged by this line at very reasonable rates, and offer a great temptation. Sailings from Glasgow, every Wednesday and Saturday; and from Moville (Londonderry—Ireland), the days following.

LINE FROM BREST AND HAVRE TO NEW YORK.

French Line. (*Compagnie Générale Transatlantique.*)

Always among the most popular lines, from its commencement, but temporarily obstructed by the late war, the French is adding widely to its vessels and influence, and inviting passage Americaward from the Continent, with excellent success. Those favorite vessels, the "Pereire," "Ville de Paris,"

"St. Laurent," "Europe," "Washington," etc., continue to supply the line to New York, with all the former luxuries; while a glance at the official announcement will show the connection to other ports, literally world-wide, and as popular and satisfactory as extensive.

LINE FROM BREMEN AND SOUTHAMPTON TO NEW YORK.

Bremen Line. (*North German Lloyd.*)

This line has a noble fleet of first-class ships in the "Rhein," "Main," "Donau," "Weser," and a score of other fine vessels, fast and reliable—not only to New York, but to Baltimore, New Orleans, and Aspinwall. Sailings from Bremen to New York, twice a week, touching at Southampton, (Eng.) and affording choice passage from London direct.

BANKING-HOUSES.

Exchange may be bought, Letters of Credit taken, and all other financial business connected with American tours and mercantile operations, safely and profitably transacted as well as many of the details of correspondence and requirement abroad—with

Duncan, Sherman & Co., cor. Pine and Nassau Sts., New York, a very old and reliable Anglo-American banking-house, through the Union Bank of London and other European agencies;

Brown Brothers & Co., 59 Wall St., another Anglo-American house of the first reputation, through Brown, Shipley & Co., Lothbury, London, and Chapel Street, Liverpool;

Jay Cooke & Co., cor. Wall and Nassau Sts., New York, American financiers of wide reputation, through Jay Cooke, McCulloch & Co., Lombard St., London;

Morton, Bliss & Co., 30 Broad St., New York, thoroughly-reliable international bankers—through Morton, Rose & Co., Bartholomew Lane, London;

Bowles, Brothers & Co., 19 William St., New York, long known as favorite American bankers in Paris—through that firm, 449 Strand, Charing Cross, London, or 12 Rue de la Paix, Paris;

Henry Clews & Co., 32 Wall St., New York—through Clews, Habicht & Co., 11 Old Broad St.,

London, now the financial agents of the United States government, for Europe;

Wells, Fargo & Co., 84 Broadway, New York, well-known bankers and express forwarders—through the same firm, 61 King William St., London;

Williams & Guion, 63 Wall St., New York—bankers, and of the well-known steamship line of the same name, before mentioned—through Alex. S. Petrie & Co., London;

John Munroe & Co., 8 Wall St., New York—through Monroe & Co., 7 Rue Scribe, Paris, favorite American house in that city.

REMINDERS TO RAMBLERS.

[See announcement cards, following, for many particulars of value and interest.]

SPECIAL ROUTES FOR TRAVELERS.

The Pennsylvania Central Railroad, now extending (by the New Jersey road to Philadelphia) from New York to Philadelphia, Harrisburg, Pittsburg, Cincinnati, etc., to Chicago and the Great West; with the especial features of the magnificent crossing of the Alleghany Mountains, and of being not only one of the most extensive in its connections, but one of the most safely and ably managed of American railways.

The Vermont Central Railroad, and leased lines connecting, furnishing among the very best and quickest of routes from both New York and Boston to the Vermont cities, Lake Champlain, Northern New York, the White Mountains, Montreal, etc., passing through some of the most picturesque mountain, river and valley scenery of the Northern States.

The Chicago and North-Western Railroad, one of the great enterprizes of the North-West, leading from Chicago, by Fulton, Clinton, Cedar Rapids, Boone, etc., to Council Bluffs, Omaha, and connecting there with the Pacific roads for California;

also with lines leading directly from Chicago by Kenosha, Racine, etc., to Milwaukie; and to Fort Howard, Green Bay, or direct connection to St. Paul and Lake Superior.

The Union and Central Pacific Railroads, direct from Council Bluffs and Omaha to San Francisco and the other California cities, and natural curiosities; by Cheyenne, Ogden, etc., with connections to Denver, the Colorado Mountains, Salt Lake City, the great Mining Regions, and the innumerable points of interest of this largest and most notable of railway-routes in the world—as well as the Pacific and its ports, and (by steamers from San Francisco) to Japan, China, the East Indies, Australia, and "round the world."

The Pacific Mail Steamship route, from New York by Kingston (Jamaica), to Aspinwall, the Panama Railroad, Panama, and steamers up the Pacific coast, to San Francisco—with connections to all ports of Central and South America, and a tropical experience not otherwise attainable anywhere to the same advantage.

The Day Line of Steamers on the Hudson River, from New York to West Point, the Catskill Mountains, etc., to Albany, supplying two of the fastest and most commodious steamers in the world, in the "Chauncey Vibbard" and "Daniel Drew," carrying music on all trips, making the whole run, in either direction, by day-light, and affording opportunities for observation of the whole line of Hudson

River scenery, unattainable by any other route or mode. [See map of the Hudson, accompanying announcement.]

The People's Line of Steamers between New York and Albany; leaving either place at evening and arriving at the other in the morning, in time for all railway connections; and supplying, in the "St. John" and "Drew," the two noblest and most luxurious specimens of inland marine architecture on the globe, with accommodations of perfect completeness and princely splendor, while still making no heavy draft on the purse.

The Hancox Line of Steamers between New York and Albany and Troy; leaving at evening and arriving in the morning, in time for all railway connections; and supplying the staunch, well-fitted and commodious boats, the "Connecticut" and "Vanderbilt," while making a specialty of affording this transit at the lowest of charges.

The Mary Powell (steamer), running as an afternoon boat from New York to West Point, Newburg, and other points on the Hudson, to Rondout (Kingston), passing through the Highlands by daylight, and affording the most charming of views of that splendid river-scenery; the boat herself a favorite and a celebrity, and well known to many Englishmen who have never visited America, from the fine picture of her, as a type of American river-steamers, in the Inman steamship office at Liverpool.

NEW YORK NOTABILITIES.

The magnificent buildings of the *Equitable Life Assurance Society*, corner of Broadway and Cedar Street, and the *Mutual Life Insurance Company*, corner of Broadway and Liberty Street—two of the actual sights of the city, without and within; while the two companies stand confessedly at the head of American Life Insurance, both in the extent of their operations and the thorough reliability of the investments made in them and by them. Features added during the past year to both, in the splendid group of statuary (by J. Q. A. Ward), over the portico of the Equitable, and the increased height and noble clock-tower of the Mutual—make the two buildings even more marked specialties of New York, than they have before been, though widely celebrated.

The office of the *Hanover Fire Insurance Company*, in the Equitable Building, Broadway and Cedar Street—perhaps the handsomest, in its fitting-up, of any in the city, and certainly among the most respectable and reliable of all, in the detail of property-security supplied by it.

The *Travelers' Insurance Company*, of Hartford, Connecticut, of which the office is at Broadway and Fulton Street; and which is doing a work of unequalled usefulness, in its insurances against every kind of accident, to which the traveler is especially liable, while even the stay-at-home by no means always escape corresponding casualties. Also a Life Assurance Company of prominence.

The diamond and fine-jewelry house of *Stevens & Co.*, 859 Broadway (near Union Square), over the door of which the magnificent Winged Lion of Venice appropriately holds its place, from the fact that the very richest and most eclectic of those works in gold, silver and precious stones, for which the Venetians and Florentines have so long been famous, are supplied at this house as at no other in America.

The *Gilsey House*, Broadway and Twenty-ninth Street, one of the noblest in the city, both in external architecture and internal arrangement, with passenger-lift, noble halls, and luxurious suits of apartments, and commending itself to travelers, of either continent, as a type of that world-wide celebrity, the American hotel of the first-class. In connection, the *Grand Union Hotel*, at Saratoga—as see following.

The *Brevoort House*, Fifth Avenue and Eighth Street, one of the most charmingly located of New York hotels, combining convenience and first fashion with quietness, and always a favorite with European tourists and those traveled Americans who have enjoyed the widest experience and know a true hotel from a caravanserai.

The *Fifth Avenue Theatre*, Twenty-fourth Street, near Fifth Avenue, one of the youngest of the first-class houses of the city, but among the most distinguished for original and successful productions (many of them from the facile pen of manager Augustin Daly), for the great average excellence of its comedy company, its vivacity of entertainment, and the fashionable character of its audiences.

The *Olympic Theatre*, Broadway near Bleecker Street, scene of many of the best triumphs of Miss Laura Keene, Mrs. John Wood, Mr. Sothern, Mr. Jefferson. etc.; but of late the acknowledged and unequalled home of pantomime, and the spot where the presence of Mr. George L. Fox, dramatic author, and the best pantomimist since Gabriel Ravel, always ensures a carnival of merriment.

Brady's Photographic and National Portrait Gallery, Broadway and Tenth Street, where the most extensive and valuable collection of the portraits of American celebrities may be found, attainable on the Continent—and at some near day to become the property of the nation, under purchase by Congress. (Also at Washington).

The *United States Watch Company* (Giles, Wales & Co.,) Maiden Lane near Broadway, with an extensive factory at Marion, New Jersey (near Jersey City), and supplying time-pieces never excelled by London or Copenhagen, at prices immeasurably below either.

The celebrated restaurant of *Nash & Fuller* (formerly the equally-celebrated "Crook & Duff's,") in the Times Building, at Printing House Square, which may be cited as the type establishment of its class in America, in extent and excellence—the resort of many celebrities, and with a popular Billiard-Hall in connection, habitually patronized by the best masters and lovers of that important branch of " mathematical science."

MISCELLANEOUS

Bonds of the *Northern Pacific Railroad*, offering most profitable investments, in connection with the great enterprise now being pushed rapidly forward to completion, and destined to unite the whole Canadian and Lake Superior region with the Pacific at Portland, Oregon (See Map).

The *Galaxy*, one of the leading magazines of America, employing many of the best pens on both sides of the Atlantic, and with specialties making it a national feature.

The *Travelers' Official Railway Guide*, published monthly by the National Railway Publication Company, Philadelphia; very correct, careful and comprehensive, and incomparably the best authority on the Western Continent, as to times of trains, connections, and every description of minute information necessary for the traveler. [To be purchased, at all news and periodical establishments.]

The *Grand Union Hotel*, Saratoga Springs, one of the largest in the world, second to no other in the extent and perfection of its appointments, with grounds of unequalled magnificence (including the Opera House—grand music and ball room of the town); and certain to win even additional celebrity under the new proprietorship, which gives it an old Saratoga-manager, in charge, and connects it with Gilsey House, New York.

The *Catskill Mountain House*, elsewhere referred to, in connection with those mountains, as one of those special points in American touring that literally must be visited, for the sake of the scenery it monopolizes, as well as for its own merits.

The *Laurel House*, Catskill Mountains, having peculiar attractions to those visiting the famed Kanterskill Falls, and many charms as a place of summer sojourn.

The *Rossin House*, Toronto, Canada, one of the largest and most complete houses in the Dominion, and deservedly a favorite with all who have occasion to visit that beautiful city of Lake Ontario.

The *Grant Locomotive Works*, Paterson, New Jersey (on the Erie Road), where the tourist can inspect one of the largest and most complete locomotive shops in the world, from which emanated the splendid engine "America," taking the great gold medal at the Paris Exhibition, and also make himself acquainted with the general appearance and construction of American locomotives.

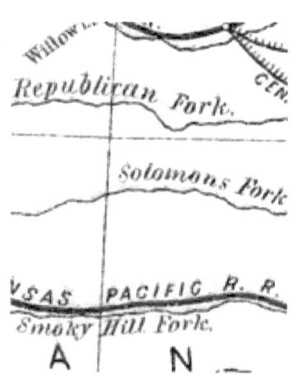

ANNOUNCEMENTS.

SHORT-TRIP GUIDE.—ANNOUNCEMENTS.

1872. CUNARD LINE. 1872.

BRITISH AND NORTH AMERICAN
ROYAL MAIL STEAMSHIPS,

BETWEEN
LIVERPOOL, BOSTON, AND NEW YORK,

CALLING AT CORK HARBOR.

SCOTIA,	*RUSSIA,*	*JAVA,*	*CUBA,*
BOTHNIA,	*SCYTHIA,*	*CHINA,*	*ABYSSINIA,*
ALGERIA,	*PARTHIA,*	*CALABRIA,*	*BATAVIA,*
SAMARIA,	*SIBERIA,*	*TARIFA,*	*TRIPOLI,*
ALEPPO,	*ATLAS,*	*SIDON,*	*PALMYRA,*
OLYMPUS,	*MARATHON,*	*MOROCCO,*	*MALTA,*
	HECLA,	*KEDAR,*	

From Liverpool—Tuesdays, Saturdays and Thursdays; calling at Cork Harbor the following days. *From New York*—Wednesdays and Saturdays. *From Boston*—Tuesdays.

Rates of Cabin Passage Money: 15 Guineas, 17 Guineas, and 21 Guineas, according to the accommodation.
Return Tickets (available for Six Months), 30 Guineas.
Rates of Passage Money by the Steamers carrying no Steerage Passengers: Chief Cabin, Twenty-six Pounds;
Second Cabin, Eighteen Pounds.
Return Tickets (available for Six Months), Chief Cabin, Fifty Pounds.

CHILDREN BETWEEN TWO AND TWELVE YEARS, HALF-FARE.

These rates include Steward's Fee and Provisions, but without Wines or Liquors, which can be obtained on board.

Passengers booked through to San Francisco, China, Japan, India, New Zealand, and Australia, by Pacific Railway and Mail Steamers.

The Passengers and Goods for New York are intended to be landed at Jersey City, within the jurisdiction of the Custom House of New York.

☞ Apply at the Company's Office, New York, to CHARLES G. FRANCKLYN, Agent; at the Company's Office, Boston, to JAMES ALEXANDER, Agent; in Halifax, to WILLIAM CUNARD; in Havre, to BURNS & MAC IVER, 21 Quai d'Orleans; in Paris, to BURNS & MAC IVER, 12 Place de la Bourse; in London, to ———— ————, 6 St. Helens Place, Bishopsgate Street; in Dundee, to G. & J. BURNS, Baltic Street; in Glasgow, to G. & J. BURNS, 30 Jamaica Street; in Belfast, to A. G. S. McCULLOCH; in Queenstown, to D. & C. MAC IVER; or to

D. & C. MAC IVER,
8 Water Street, Liverpool

INMAN LINE.

NEW YORK TO LIVERPOOL
TWICE EVERY WEEK,
(From Pier 45 North River, New York.)

Saturday Line.	Thursday Line.
CITY OF CHESTER.	CITY OF NEW YORK.
CITY OF RICHMOND.	CITY OF LONDON.
CITY OF MONTREAL.	CITY OF WASHINGTON.
CITY OF BRUSSELS.	CITY OF BALTIMORE.
CITY OF PARIS.	CITY OF ANTWERP.
CITY OF BROOKLYN.	CITY OF BRISTOL.

RATES OF PASSAGE.

To Liverpool, $75 and $90, gold. | From Liverpool, $75 and $90 gold. (15 to 18 Guineas.)

Round Trip Tickets, $135 and $150, gold.

Steamers leave LIVERPOOL every Tuesday and Thursday.
" " QUEENSTOWN, every Wednesday & Friday.
" " NEW YORK, every Thursday & Saturday.

For further information, apply at the Company's Offices:

Liverpool, WM. INMAN, 62 *and* 63 *Tower Buildings.*
Queenstown, C. & W. D. SEYMOUR.
London, EIVES & ALLEN, 61 *King William Street.*
Paris, BOWLFS BROS. & CO., 12 *Rue de la Paix.*
" J. W. TUCKER & CO., 3 *and* 5 *Rue Scribe.*
" JULES DECOUÉ, 48 *Rue Notre Dame des Victoires, Place de la Bourse.*
Boston, M. S. CREAGH, 102 *State Street.*
Philadelphia, O'DONNELL & FAULK, 402 *Chestnut Street.*
Chicago, FRANCIS C. BROWN, 39 *W. Kinzie St*, and in
New York to

JOHN G. DALE, *Agent,*
15 BROADWAY.

SHORT-TRIP GUIDE.—ANNOUNCEMENTS.

LIVERPOOL
AND
NEW YORK STEAMERS

Carrying the United States Mails.

MANHATTAN,	NEBRASKA,
IDAHO,	
MINNESOTA,	NEVADA,
WISCONSIN,	WYOMING,
DACOTAH,	MONTANA.

The above Steamers are New, of the Largest Class, and built expressly for the Trade. Have five Water-Tight Bulkheads, and carry experienced Officers, Surgeons and Stewardesses.

The Saloon accommodations and attendance are unsurpassed by any Atlantic Steamers.

SAILING FROM

Liverpool and New York on Wednesdays,

(Calling at Queenstown to land and receive Mails and Passengers.)

PASSAGE.

New York to Liverpool . . . $80 gold.
Liverpool to New York . £15 or £18.
(According to Staterooms.)

AGENTS,

GUION & CO.,	**WILLIAMS & GUION,**
Liverpool.	*63 Wall Street, New York*
A. S. PETRIE & CO.,	**J. M. CURRIE,**
11 Old Broad St., London.	*Paris and Havre.*

SHORT-TRIP GUIDE.—ANNOUNCEMENTS.

NATIONAL LINE.

Steamers Weekly, between

NEW YORK, LIVERPOOL AND QUEENSTOWN.

SPAIN,	- 4,871 tons.	ENGLAND,	-	3,441 tons.
EGYPT,	5,150 "	THE QUEEN,		4,470 "
ITALY,	- 4,340 "	HELVETIA,	-	4,020 "
HOLLAND,	3,847 "	ERIN,	-	4,030 "
FRANCE,	- 3,676 "	CANADA,	-	4,500 "
DENMARK,	3,724 "	GREECE,	-	4,500 "

The above powerful British-built Iron Steamships, with spar decks, and water-tight compartments, the largest in the trade, during the season of 1872, will form this favorite line, leaving

LIVERPOOL, - - EVERY WEDNESDAY.
QUEENSTOWN, - - EVERY THURSDAY.
NEW YORK, - - - EVERY SATURDAY.

From the Company's Wharves,

Piers 44 and 47 North River.

RATES OF PASSAGE PAYABLE IN U. S. CURRENCY:

	1st.	2d
To Liverpool or Queenstown,	$75	$65
London,	85	75
Hamburg,	100	90
Bremen,	110	100
Antwerp,	100	90
Havre,	100	90
Paris,	100	90
Tickets to Liverpool and return,	130	
Prepaid Cabin Tickets from Liverpool or Queenstown,	75	65

The 2d rate includes first-class to London, Paris, &c.

F. W. J. HURST, Manager,

69 BROADWAY.

NEW YORK AND LIVERPOOL
CALLING AT QUEENSTOWN.

Sailing from New York on Saturdays, from Liverpool on Thursdays.

Average Passage about Nine Days.

| OCEANIC, | BALTIC, | CELTIC, |
| ATLANTIC, | REPUBLIC, | ADRIATIC. |

The Six Largest Steamships afloat.

Those intending to cross the Atlantic would do well to inspect the accommodations offered by these new and magnificent vessels. Nothing has been left undone to promote the comfort and convenience of passengers, and to make the voyage agreeable. *Pianos* and *Libraries* have been provided; and Main Saloons, State Rooms, Hot and Cold Baths, Barber's Shops, &c., are situated in the midship sections, where least motion is felt.

Surgeons and Experienced Stewardesses accompany these Boats

Rates:—Saloon, $80 gold; Excursion, $140 gold; Steerage, Outwards, $30 currency.

WHITE STAR LINE OFFICES, Messrs. Ismay, Imrie & Co.,
10 *Water Street, Liverpool.*

J. H. SPARKS, Agent,
19 *Broadway, New York.*

1872.
STEAMERS TO FRANCE DIRECT.

Transit by Railroad, and crossing the English Channel avoided.

The General Transatlantic Co's
FIRST-CLASS STEAMSHIPS

PEREIRE.	ATLANTIC.	GUADELOUPE
VILLE DE PARIS.	FRANCE.	DESIRADE.
ST. LAURENT.	PANAMA.	GUYANE.
VILLE DU HAVRE.	VILLE DE ST. NAZAIRE.	SONORA.
EUROPE.	VILLE DE BORDEAUX.	CARIABE.
WASHINGTON.	LOUISIANE.	CACIQUE.
VILLE DE BREST.	FLORIDE.	CARANELLE.
NOUVEAU-MONDE.	MARTINIQUE.	

Postal Lines of the General Transatlantic Co.

From HAVRE to NEW YORK, calling at Brest, and vice versa, . . . *Saturdays, Twice a Month.*

From ST. NAZAIRE to VERA CRUZ, calling at Santander, St. Thomas and Havana, and vice versa, ... *Once a Month.*

From ST. NAZAIRE to ASPINWALL, calling at Martinique, La Guayra and St. Martha, and vice versa, ... *Once a Month.*

From PANAMA to VALPARAISO, calling at Intermediate Ports, and vice versa, . . *Once a Month.*

BRANCH LINES.

From ST. THOMAS to ASPINWALL, calling at Porto Rico, Hayti, Santiago de Cuba, Kingston, Jamaica, and vice versa, . . . *Once a Month.*

From ST. THOMAS to FORT DE FRANCE, (Martinique,) calling at Basse Terre, (Guadeloupe,) Pointe a Petre, (Guadeloupe,) St. Pierre, (Martinique) and vice versa, . . . *Once a Month.*

From FORT DE FRANCE, (Martinique.) to CAYENNE, calling at St. Lucia, St. Vincent, Grenada, Trinidad, Demerara, Surinam, and vice versa, ... *Once a Month.*

The splendid Steamers of the South Pacific Line leave Panama for Valparaiso and Intermediate Ports, on the 30th of every month, and connect closely with the Steamers of the Pacific Mail S. S. Company leaving New York on the 15th of every month for Aspinwall.

For Rates of Passage and Freight, Dates of Departure, or further information, apply to

GEO. MACKENZIE, Agent, 58 Broadway.

ANCHOR LINE.

Between New York and Glasgow,

Sailing Every Wednesday and Saturday.

The Powerful Clyde-Built Steamers,

BOLIVIA, (new.)	AUSTRALIA.
UTOPIA, "	CALIFORNIA.
VICTORIA, "	EUROPA.
ANGLIA.	INDIA.
SCANDINAVIA.	COLUMBIA.
ITALIA, (new.)	OLYMPIA.

And more than a score of other First-Class Ships, comprising one of the largest fleets in any service.

Passengers booked to or from Liverpool, Glasgow, London, Queenstown, or Londonderry, at as low rates as by any other first-class line.

Through Tickets issued to and from any Seaport or Railway Station in the World.

The Anchor Line Steamers are

FIRST-CLASS IN EVERY RESPECT,

Safe, Comfortable, Reliable, Splendidly Equipped, and in their Appointments and Equipments, not excelled by any other line.

COMPANY'S OFFICES:

LIVERPOOL, 17 *Water St.*   *GLASGOW*, 51 *Union St.*
LONDONDERRY, 96½ *Foyle St.*   *CHICAGO*, 324 *Wabash Av.*
NEW YORK, 7 *Bowling Green.*

HENDERSON BROTHERS, Agents.

NORTH GERMAN LLOYD.

STEAM BETWEEN

Bremen, (via Southampton and Havre,)

AND THE PORTS OF

New York, Baltimore, New Orleans, Havana, Aspinwall, &c.

The Screw Steamers of the North German Lloyd:

RHEIN,	WESER,	AMERICA,
MAIN,	HERMANN,	BREMEN,
DONAU,	NEW YORK,	HANSA,
DEUTSCHLAND,	HANNOVER,	FRANKFURT,
KOLN,	STRASSBURG,	MOSEL,
NECKAR,	BALTIMORE,	BERLIN,
LEIPZIG,	OHIO,	BISMARCK,
KONIG WILHELM I.	KRONPRINZ FRIEDRICH WILHELM.	
GENERAL VON ROON.		GRAF MOLTKE.

These Vessels carry the German, British and United States Mails, and leave

BREMEN for New York, every Wednesday and Saturday.
    " for Baltimore, on alternate Wednesdays.
    " for New Orleans, once every week.
    " for Aspinwall,   ' " "
NEW YORK for Bremen, via Southampton, every Wednesday and Saturday.
BALTIMORE for Bremen, on alternate Wednesdays.
NEW ORLEANS for Bremen, once every week.
ASPINWALL for Bremen,   " " "

The above vessels have been constructed in the most approved manner; they are of 3,000 tons, and 700 horse-power each, and are commanded by men of character and experience, who will make every exertion to promote the comfort and convenience of passengers. They touch at Southampton, on the outward trip, for the purpose of landing passengers for England and France.

These Vessels take Freight to Bremen, London, Hull, Rotterdam, Antwerp, and Hamburg, for which through bills of lading are signed.

An experienced surgeon is attached to each vessel.

All letters must pass through the post office

Specie taken to Havre, Southampton and Bremen at the lowest rates.

For Prices of Passage, and all further particulars, apply to NORTH GERMAN LLOYD, *Bremen;* KELLER, WALLIS & POSTLETHWAITE, *Southampton;* PHILLIPPS, GRAVES, PHILLIPPS & CO., *London;* LHERBETTE, KANE & CO., *Havre and Paris.* OELRICHS & CO., *New York;* A. SCHUMACHER & CO., *Baltimore;* ED. T. STOCKMEYER, *New Orleans;* H. HYMAN & CO., *Havana;* W. P. MAAL Y HERMANO, *Aspinwall*

Brown Brothers & Co.,

59 *Wall Street, New York.*

BILLS OF EXCHANGE on Great Britain and Ireland.

COMMERCIAL AND TRAVELING CREDITS issued.
available in any part of the world.

TELEGRAPHIC TRANSFERS OF MONEY
to and from London and Liverpool.

ADVANCES
made on Cotton and other produce.

BROWN, SHIPLEY & CO.,
Founder's Court, Lothbury, London.

BROWN, SHIPLEY & CO.,
Chapel Street, Liverpool.

Morton, Bliss & Co.,
BANKERS.

30 *Broad Street, New York,*

ISSUE
CIRCULAR NOTES AND LETTERS OF CREDIT
for Travelers; also
COMMERCIAL CREDITS
available in all parts of the world.

Negotiate First-Class RAILWAY, CITY and STATE LOANS,

Make TELEGRAPHIC TRANSFERS OF MONEY,

Allow INTEREST ON DEPOSITS, and

Draw EXCHANGE on

MORTON, ROSE & CO., London.

HOTTINGUER & CO., Paris.

HOPE & CO., Amsterdam.

SHORT-TRIP GUIDE.—ANNOUNCEMENTS.

Duncan, Sherman & Co.,
BANKERS,

Corner of Pine and Nassau Sts., New York,

ISSUE

CIRCULAR NOTES AND TRAVELING CREDITS,

Available in all the Principal Cities of the World.

TRANSFERS OF MONEY BY TELEGRAPH TO EUROPE, CUBA, AND THE PACIFIC COAST.

Accounts of Country Banks and others Received.

John Munroe & Company,
BANKERS,

No. 8 Wall Street, New York, and
No. 41 State Street, Boston,

ISSUE

CIRCULAR LETTERS OF CREDIT FOR TRAVELERS,

ON THE

CONSOLIDATED BANK, London,

AND ON

Munroe & Company,

No. 7 Rue Scribe, PARIS.

EXCHANGE ON LONDON AND PARIS.

WELLS, FARGO & COMPANY,
BANKERS,

AND EXPRESS FORWARDERS TO ALL PARTS OF THE WORLD.

PRINCIPAL OFFICES:

84 *BROADWAY, New York.*
94 *WASHINGTON ST., Boston.*
MONTGOMERY & CALIFORNIA STREETS, San Francisco,
61 *KING WILLIAM ST., London.*

WITH AGENCIES IN PARIS, BREMEN,
And all the Principal Cities and Towns in the United States and Territories.

DOMESTIC AND FOREIGN EXCHANGE,

AND

Telegraphic Transfers for sale.

LETTERS OF CREDIT ISSUED TO TRAVELLERS.

Interest allowed on Deposit Accounts.

Particular attention given to arranging
TRAVELING CREDITS IN THE WESTERN STATES.

BOWLES BROTHERS & Co.,

PARIS, 12 *Rue de la Paix,*
LONDON, 449 *Strand, Charing Cross,*
NEW YORK, 19 *William Street,*
BOSTON, 27 *State Street.*

ISSUE

BILLS ON PARIS AND LONDON,

In sums to suit: also,

CIRCULAR LETTERS OF CREDIT,

AVAILABLE IN ALL THE CITIES OF EUROPE.

Letters addressed to our care, receive most careful attention, each being registered at our office on receipt and delivery.

CORRESPONDENTS OF THE FOLLOWING BANKS:

The Union Bank of London.	The Oriental Bank Corporation.
Messrs. J. S. Morgan & Co.	The National Bank of Scotland.
The Bank of California.	Messrs. Wells, Fargo & Co.

Munster Bank and Branches, Ireland.

Banking House of HENRY CLEWS & CO.,

32 WALL STREET, NEW YORK.

Circular Notes and Letters of Credit for Travelers,

ALSO,

COMMERCIAL CREDITS ISSUED,

Available throughout the world.

Bills of Exchange and Telegraphic Transfers

Of Money on Europe, San Francisco, and the West Indies.

Deposit Accounts Received, subject to Check at sight. Interest allowed on all Daily Balances.

Government, State, City and Railroad Loans Negotiated.

CLEW'S, HABICHT & CO.,

11 OLD BROAD ST., LONDON,

Bankers, and Fiscal Agents of the United States Government at London, for all Foreign Countries.

WILLIAMS & GUION,

63 WALL ST., NEW YORK.

Travelers' and Commercial Credits Issued,

Available in all parts of Europe, &c.

BILLS OF EXCHANGE

Drawn in sums to suit purchasers; ALSO CABLE TRANSFERS.

Advances Made upon Consignments of Cotton and other Produce to Ourselves or Correspondents.

GUION & CO.,
LIVERPOOL.

ALEX. S. PETRIE & CO.,
LONDON.

Jay Cooke, McCulloch & Co.

41 LOMBARD STREET, 41
LONDON.

JAY COOKE & CO.

20 WALL ST.,   114 SO. THIRD ST.,
NEW YORK.   *PHILADELPHIA.*

15th STREET, opp. U. S. Treasury,
WASHINGTON.

Exchange Sold on all Leading Cities
OF
UNITED STATES AND CANADA,
PAYABLE IN DOLLARS, GOLD, or CURRENCY.

Sterling Drafts & Cable Transfers on America.

CIRCULAR LETTERS OF CREDIT
For Travellers.

COMMERCIAL CREDITS.

SHORT-TRIP GUIDE.—ANNOUNCEMENTS.

HANOVER
FIRE INSURANCE COMPANY,
OF THE CITY OF NEW YORK,

Office, No. 120 Broadway, cor. Cedar St.,

(EQUITABLE LIFE ASSURANCE CO'S BUILDING.)

INCORPORATED 1852.

B. S. WALCOTT, *President.*
I. REMSEN LANE, *Secretary.*
HENRY KIP, *Assistant Secretary.*

Cash Capital,	$400,000 00
Cash Assets,	$872,627 91

AGENCIES IN ALL THE PRINCIPAL TOWNS IN THE UNITED STATES.

Eastern Agency Department, - - THOMAS JAMES, *Actuary.*
Western and Southern Agency Department, "The Underwriters Agency," A. STODDART, *General Agent.*

The Mutual Life

Insurance Company

OF NEW YORK,

144 AND 146 Broadway,

NEW YORK CITY.

F. S. WINSTON, *President.*

Cash Assets OVER $51,000,000.

Invested in Loans on Bond and Mortgage, or United States Stocks.

Issues every approved description of Life and Endowment Policies on selected lives at Moderate Rates, returning all surplus annually to the policyholders, to be used either in payment of premiums, or to purchase additional insurance, at the option of the assured.

RICHARD A. McCURDY, *Vice-President.*
JOHN M. STUART, *Secretary.*
WM. H. C. BARTLETT, *Ass't Secretary.*
F. SCHROEDER, *Actuary.*
LEWIS C. LAWTON, *Ass't Actuary.*

SHORT TRIP GUIDE.—ANNOUNCEMENTS.

THE
EQUITABLE
LIFE ASSURANCE SOCIETY
OF THE UNITED STATES,

No. 120 Broadway, New York.

ASSETS.	$18,000,000 00
INCOME, ...	8,000,000 00
SUM ASSURED, (New Business) 1871,	41,300,000 00

ALL CASH.

PURELY MUTUAL.   ANNUAL DIVIDENDS.

The New Business of the Equitable is larger than that of any other Life Insurance Company in America or Europe.

The average annual growth of the Society's Permanent Business—Risks in Force—since its organization, has been greater than that of any other leading Company.

Its average percentage of "*Losses*" to "*Amount in Force*," during the last five years, is less than that of any other of the older and larger Companies of the United States.

Its "*Expenses*," compared with "*Income*" are much less than the average of all other New York Companies.

OFFICERS:

WILLIAM C. ALEXANDER, President.
HENRY B. HYDE, Vice-President.
JAMES W. ALEXANDER, 2d Vice-President.
SAMUEL BORROWE, Secretary.
WILLIAM ALEXANDER, Ass't Secretary.
GEORGE W. PHILLIPS, Actuary.
WILLIAM P. HALSTED, Auditor.

IMPORTANT  TO TRAVELERS.

THE TRAVELERS INSURANCE CO.
OF HARTFORD, CONN.
CASH ASSETS, $2,000,000.

Grants everything desirable in

LIFE AND ACCIDENT INSURANCE,
ON THE MOST FAVORABLE TERMS.

ACCIDENT DEPARTMENT.

The TRAVELERS INSURANCE COMPANY, in its Accident Department, is a General Accident Insurance Company granting policies of insurance against Death or wholly Disabling Injury by ACCIDENT, to Men of all trades, professions and occupations, at rates within the reach of all. Policies are written for a term of one to twelve months each, and insure a sum of $500 to $10,000, at rates of premiums designated to cover risks at home and abroad—and covering all varieties of occupations.

LIFE DEPARTMENT.

In its Life Department, the TRAVELERS grants full LIFE and ENDOWMENT Policies, embracing the best features of the best companies, as to non-forfeiture, terms of payment, etc., but without any of the complications or uncertainties of the note system.

All policies non-forfeitable. Its five, ten, fifteen and twenty year policies can be converted into endowments, at the option of the insurant.

This feature is original with this Company.

THE TRAVELERS furnishes everything desirable in either *Life* or *Accident* insurance. It has issued 300,000 general accident policies, and *paid fifteen thousand claims* for death or injury by accident. It has issued between ten and twelve thousand full life policies, since the Life Department was established, and is making good and safe progress as a life company. Its capital and surplus amount to $1,850,000—giving $182 cash assets for every $100 of liability, thus furnishing an amount of financial security rarely, if ever, equalled by any Life Insurance Company.

JAMES G. BATTERSON, *President.*
RODNEY DENNIS, *Secretary.*
CHAS. E. WILSON, *Ass't Secretary.*
GEO. B. LESTER, *Actuary.*

POLICIES WRITTEN IMMEDIATELY ON APPLICATION, AT THE

New York Office, 207 *Broadway.*

R. M. JOHNSON, Manager.

United States Watch Company.

GILES, WALES & CO., Marion, N. J.
GILES, WALES & CO., No. 13 Maiden Lane,
NEW YORK.

Testimonial Records of Performances
OF
MARION UNITED STATES WATCH CO'S WATCHES,

Admitted to be Unparalleled in the Trade.

Watch No. 1089—variation, 2 seconds in 14 months.
  L. E. CHITTENDEN, late Register U. S. Treasury.
Watch No. 1124—variation, 6 seconds in 7 months.
  A. L. DENNIS, President N. J. R. R. & T. Co.
Watch No 1037—variation, 5 seconds per month.
  HENRY SMITH, Treasurer Panama R. R. Co., N. Y.
Watch No. 2617—variation, 15 seconds in 12 months.
  I. VROOMAN, Engineer N. Y. C. & H. R. R.
Watch No. 4026—variation, 3 seconds in 2 months.
  JOSHUA I. BRAGG, Conductor N. J. R. R.
Watch No. 24,008—variation, 6 seconds in 5 months.
  CHARLES H. WOLF, Pearl Street, Cincinnati, Ohio.
Watch No. 1143—variation, 39 seconds in 8 months.
  JAMES B. RYER, of Kelty & Co., 722 Broadway, N. Y.
Watch No. 1894—variation, 8 seconds in 6 months.
  H. COTTRELL, 123 Front Street, N. Y.
Watch No. 1205—variation, 7 seconds in 11 months.
  A. H. KING, Vice-Pres't N. J. Car Spring Co., N. Y.
Watch No. 1788—variation, 20 seconds in 5½ months,
  by Greenwich Observatory Time, London.
  HENRY MORFORD, Author-Prop'r Short-Trip Guide, N.Y.

Best Investments for 1872.

NORTHERN PACIFIC RAILROAD BONDS.

The rapid progress of the Northern Pacific Railroad toward completion, is shown in the fact that to the Missouri River the Road is under contract to be finished during the present year.

This will give to its bondholders a lien upon millions of acres of the finest land in the country, besides their mortgage on the road and its earnings, thus making doubly secure the investment, which, in all its features, is one of the safest and most reliable next to the Government loans. The redemption of 5-20s by the Treasury indicates low rates of interest to the public creditors hereafter; and as the calling-in is now rapid, we strongly recommend to the holders thereof an immediate exchange for the 7-30 GOLD BONDS of the NORTHERN PACIFIC RAILROAD.

They are free of United States Tax, and are issued in the following denominations: Coupons, $100, $500, and $1,000; Registered, $100, $500, $1,000, $5,000 and $10,000.

The bonds, which are being rapidly sold, are secured by a first and only mortgage on over two thousand Miles of Road with rolling stock, buildings, and all other equipments, and on over TWENTY-THREE THOUSAND ACRES of Land to every mile of finished road. This land, agricultural, timbered and mineral, amounting in all to more than Fifty Million Acres, consists of alternate sections, reaching twenty to forty miles on each side of the track and extending in a broad, fertile belt from Wisconsin through the richest portions of Minnesota, Dakota, Montana, Idaho, Oregon and Washington, to Puget Sound.

All marketable securities received in exchange. Full particulars furnished by

JAY COOKE & CO.,
New York, Philadelphia and Washington.

SHORT-TRIP GUIDE.—ANNOUNCEMENTS.

The Great Trans-Continental All-Rail
Union & Central Pacific Route
Via Council Bluffs and Omaha,
is
226 Miles the SHORTEST,
FROM THE ATLANTIC COAST TO
SALT LAKE CITY,
SACRAMENTO,
SAN FRANCISCO,
AND ALL POINTS IN THE
SANDWICH ISLANDS,   JAPAN,
NEW ZEALAND,   CHINA,
AUSTRALIA,   INDIA.

Five Hours the Quickest Route
To DENVER, COL., NEW MEXICO and ARIZONA.

PULLMAN'S PALACE DAY AND SLEEPING CARS are run on all Express Trains, and passengers are *cautioned* that only those who are ticketed via Omaha are sure of securing berths to points west of Cheyenne.

Double Berth—Omaha to Ogden, $3; Ogden to San Francisco $6.

To *Tourists, Pleasure* and *Health Seekers*, this Route offers unrivaled attractions in the beautiful Platte Valley; the grand scenery of the Snowy Range; the Passage of the Rocky Mountains (8,242 feet above the sea); the Laramie Plains; the Wahsatch and Uintah Mountains; the wild and weird Echo and Weber Cañons; the Great Salt Lake and its Mormon City, surrounded by lofty mountains, rivers filled with trout, hunting grounds, medicinal springs, etc.; the Humboldt Sink; the Sierra Nevadas the Palisades; the beautiful mountain lakes of Tahoe and Donner, and the passage of Cape Horn; making a two thousand mile panorama of unequalled grandeur and beauty.

Through Tickets for sale by F. KNOWLAND, General Agent, No. 287 Broadway, New York, and at all principal ticket offices in the country. Members of Colonies and Excursion Parties should address the General Ticket Agent for rates and arrangements.

T. E. SICKLES,   **THOS L. KIMBALL,**
Gen'l Sup't, U. P. R. R., Omaha.   Gen'l Ticket Agent, U P. R. R., Omaha.

A. N. TOWNE,   **T H. GOODMAN,**
Gen'l Supt. C. P. R. R., Sacramento   G. P. A., C. P. R. R., Sacramento

[SEE NEXT PAGE.]

SHORT-TRIP GUIDE.—ANNOUNCEMENTS.

THE
Union & Central Pacific Line.
TRIP AROUND THE WORLD.

From ATLANTIC CITIES to OMAHA, (Nebraska), via the Great Trunk Lines of Railway,—about 1,400 miles, in 50 hours.

From OMAHA to SAN FRANCISCO, (California), via Union and Central Pacific Railroads.—1,914 miles, in 4 days and 6 hours.

From SAN FRANCISCO to YOKOHAMA, (Japan), by Pacific Mail Line Steamers,—4,700 miles, in 22 days.

From YOKOHAMA to HONG KONG, (China), by Pacific Mail or Peninsular and Oriental Steamers,—1,600 miles, in 6 days.

From HONG KONG to CALCUTTA, (India), by Peninsular and Oriental Steamers,—3,500 miles, in 14 days.

Or from San Francisco to Calcutta via Australia, as follows:

From SAN FRANCISCO to HONOLULU, (Sandwich Islands), by United States, New Zealand and Australia Mail Steamship Line,—2,110 miles, in 10 days.

From HONOLULU to AUCKLAND, (New Zealand), by U. S., New Zealand and Aus. S. S. Line,—3,800 miles, in 14 days.

From AUCKLAND to SYDNEY, (Australia), by U. S., New Zealand and Aus. S. S. Line,—1,277 miles, in 5 days.

From SYDNEY to MELBOURNE, (Australia), by Peninsular and Oriental Steamers,—560 miles, in 3 days.

From MELBOURNE to GALLE, (Ceylon), by Peninsular and Oriental Steamers,—4,670 miles, in 21 days.

From GALLE to CALCUTTA, (India), by Peninsular and Oriental Steamers,—1,315 miles, in 7 days.

From CALCUTTA to BOMBAY, (India), by the East Indian and Great Indian Peninsular Railways,—1,400 miles, in 2 days.

From BOMBAY to SUEZ, (Egypt), by Peninsular and Oriental Steamers,—3,600 miles, in 14 days.

From SUEZ to ALEXANDRIA, (Egypt), by Rail along the Suez Canal,—225 miles, in 12 hours.

From ALEXANDRIA to BRINDISI, (Italy), by Peninsular and Oriental Steamers,—850 miles in 3 days.

From BRINDISI to LONDON, (England), by Rail, via Paris or the Rhine,—1,200 miles, in 3 days.

From LONDON to LIVERPOOL, (England), by Railway,—200 miles, in 5 hours.

From LIVERPOOL to ATLANTIC CITIES, (America), by either of the Great Atlantic S. S. Lines,—3,000 miles, in 10 days.

Total distance, 23,589 miles. Time, 81 days.

OFFICES:

No. 287 Broadway, New York.

No. 2 New Montgomery St., San Francisco; and

H. STARR & CO'S Office, 22 Moorgate Street. London.

[SEE PRECEDING PAGE.]

THE
Pennsylvania Central Railroad,
PITTSBURGH, FORT WAYNE & CHICAGO RAILWAY,
AND
PAN-HANDLE ROUTE

Furnish the shortest and quickest route from NEW-YORK, and all Eastern Cities to

CHICAGO, MILWAUKEE, ST. PAUL, OMAHA, CINCINNATI, LOUISVILLE, NEW ORLEANS, INDIANAPOLIS, ST. LOUIS, KANSAS CITY, DENVER, SALT LAKE CITY, SAN FRANCISCO, and all points West, Northwest and Southwest.

Pullman Palace Day and Night Cars

Are run from New York to all principal Cities.

A large proportion of STEEL RAILS now in use.

Trains equipped with WESTINGHOUSE PATENT CAR BRAKE, ensuring comfort and safety.

Rates of fare always as low as by any other route.

HENRY W. GWINNER,	JOHN H. MILLER,
Gen'l Pass. & Ticket Agent,	*Gen'l Eastern Agent,*
PHILADELPHIA.	NEW YORK.

A. J. CASSATT, *Gen'l Superintendent,* PHILADELPHIA.

SHORT-TRIP GUIDE — ANNOUNCEMENTS.

Pacific Mail Steam Ship Co's

THROUGH LINE TO

California, Japan and China,

Via PANAMA and SAN FRANCISCO,

Carrying Mails, Passengers and Freight to KINGSTON, (JAMAICA,) ASPINWALL, PANAMA, and other Central American and South American Ports, and SAN FRANCISCO; and thence to YOKOHAMA, HONG-KONG, SHANGHAE, NAGASAKI and HIOGO, connecting at Hong-Kong with Steamers for Ports of the CHINA COAST and INDIA.

The Large and Splendid Steamers of this Line
Leave Pier 42, N. R., foot Canal St., N. Y.,
At Twelve o'clock, noon,
On the 15th & 30th of Every Month,
(Except when those days fall on Sunday, then on the Saturday preceding,) arriving at Aspinwall on or about the 9th and 24th of each month.

GREATLY REDUCED THROUGH-PASSAGE RATES,
NEW YORK TO SAN FRANCISCO.

CHINA LINE,

Between California, Japan and China.

Magnificent Steamers of the China Fleet leave San Francisco, 1st and 16th of every month, (except when those dates fall on Sunday, then on preceding Saturday,) for Yokohama, Hong-Kong, and connections with all Ports of Japan, China and British India.

For Passage Tickets or further information apply t
THE COMPANY'S TICKET OFFICE,
On the Wharf only, Pier 42, N. R., foot of Canal St., N. Y.
F. R. BABY, Agent.

Or, in San Francisco, to
ELDRIDGE & IRWIN, Agents.

SHORT-TRIP GUIDE.—ANNOUNCEMENTS.

The Vermont Central Railroad
AND LEASED LINES

Between New London, Montreal and Ogdensburgh.

800 *Miles of Road under one Management,* THROUGH THE MOST PICTURESQUE SCENERY IN NEW ENGLAND.

Most Direct and Popular Route between New York and Montreal, via New London, or Hudson River, or Connecticut River.

Shortest and Best Route between Boston and Montreal and Ogdensburgh, via Fitchburg or Lowell, and Favorite Route between Boston and Saratoga, via Rutland and Bellows Falls.

Direct Connection at Montreal and Ogdensburgh with Grand Trunk Railway for all Points West, and at Ogdensburgh with Vermont Central. Steamers for all points on the Great Lakes

Pullman Palace, Drawing Room, and Sleeping Cars, on all Trains.

J. GREGORY SMITH, *President.*
GYLES MERRILL, *Gen'l Sup't.*
LANSING MILLIS, *Gen'l Eastern Agent,*
65 Washington St., Boston.

CHICAGO AND NORTH WESTERN RAILWAY.

Passengers for all Points North or West of Chicago and the

PACIFIC COAST,

Will find this the Shortest and most Comfortable Route, as it is the Line over which the Celebrated

PULLMAN DINING CARS

AND

SLEEPING COACHES

ARE RUN BETWEEN

CHICAGO AND SAN FRANCISCO.

Passengers should be particular to ask for Through Tickets via

Chicago and North-Western Railway,

On sale at all Principal R. R. Offices in the U. S. and Canadas.

IN CHICAGO,

AT THE COMPANY'S DEPOTS.

H. P. STANWOOD,	M. HUGHITT,
Gen'l Ticket Agent.	Gen'l Sup't.

G. T. NUTTER,

General Eastern Agent, 229 Broadway, New York.

SHORT-TRIP GUIDE.—ANNOUNCEMENTS.

PEOPLE'S LINE
OF
STEAMERS,
BETWEEN
NEW YORK AND ALBANY.

NEW YORK TO ALBANY.

STEAMERS

St. JOHN,	DREW,
Capt. W. H. CHRISTOPHER,	Capt. S. J. ROE,
MONDAY,	*TUESDAY,*
WEDNESDAY,	*THURSDAY,*
FRIDAY.	*SATURDAY.*

FROM PIER 41, NORTH RIVER,

(Near Jersey City Ferry, Desbrosses Street),

AT 6.00 P. M,

Connecting with Trains of New York Central, Albany and Susquehanna, Rensselaer and Saratoga, and Boston and Albany Railways; and Steamers on Lake George and Lake Champlain to Burlington, White Mountains, Montreal, Quebec, &c.

☞ TICKETS can be had at the Office on the Wharf, and BAGGAGE checked to destination; also at Dodd's Express offices, 944 Broadway, New York, and No. 1 Court Street, Brooklyn.

Telegraph Office on the Wharf.

☞ Passengers leaving WASHINGTON at 8.00 A. M., BALTIMORE at 9.40 A. M., PHILADELPHIA at 1.20 P. M., arrive in NEW YORK at 4.50 P. M. in time to connect as above.

ALBANY TO NEW YORK.

STEAMERS

DREW,	St. JOHN,
Capt. S. J. ROE,	Capt. W. H. CHRISTOPHER,
MONDAY,	*TUESDAY,*
WEDNESDAY,	*THURSDAY,*
FRIDAY.	*SATURDAY.*

FROM STEAMBOAT LANDING,

AT 8.15 P. M.,

ON ARRIVAL OF TRAINS FROM NORTH AND WEST.

Baggage conveyed from N. Y. C. R. R. Depot to the Boats, FREE.

☞ TICKETS to Philadelphia, Baltimore, and Washington City, for sale on the Boats, and BAGGAGE checked to destination.

HUDSON RIVER R. R. TICKETS taken for passage, including State Room Berth,
JOHN C. HEWITT,
General Ticket Agent.

SHORT TRIP GUIDE.—ANNOUNCEMENTS.

Hudson River
BY
DAYLIGHT.

ALBANY & NEW YORK
DAY LINE
OF
STEAMBOATS,
"C. VIBBARD,"
AND
"DANIEL DREW,"

Leaving New York

From Vestry Street Pier at 8.30 A. M., and 34th St. at 8.45 A. M.,

Landing at Cozzens's, West Point, Newburgh, Poughkeepsie, Rhinebeck, Catskill, and Hudson. AFFORDING THE BEST MODE OF ENJOYING THE UNSURPASSED SCENERY, and of reaching the "Overlook" and "Catskill" Mountain Houses, Lebanon Springs (*via* Hudson), Sharon Springs by special train *via* Susquehanna Railway (all rail from Albany), Saratoga Springs, and all points north and west.

LEAVE ALBANY every morning, on arrival of Trains from Saratoga and the north, and from Sharon, etc.

ISAAC L. WELSH,
General Ticket Agent.

SHORT-TRIP GUIDE.—ANNOUNCEMENTS.

AFTERNOON BOAT
FROM
NEW YORK TO WEST POINT, NEWBURG, POUGHKEEPSIE.
RONDOUT AND KINGSTON.

THE SPLENDID AND FAVORITE STEAMER

MARY POWELL

Leaves New York, every afternoon at 3.30, from VESTRY STREET PIER, for WEST POINT, NEWBURG, POUGHKEEPSIE, RONDOUT AND KINGSTON; landing at *CORNWALL*, *MILTON*, and other popular places; passing

THROUGH THE HIGHLANDS BY DAYLIGHT,

and affording unequalled facilities for reaching the places named, and enjoying the noble scenery of the river, as well as reaching, in the most convenient manner, that charming place of summer resort in the Catskill Mountains—

THE OVERLOOK HOUSE.

Returning, leaves Rondout every morning at 5.30 A. M., reaching New York at 10.30.

1872. NEW YORK AND TROY STEAMBOAT COMPANY, 1872.

Great Reduction in Fare & State Rooms.

THE OLD ESTABLISHED LINE FOR

ALBANY AND TROY,

LANDING AT CATSKILL.

THE ELEGANT STEAMERS

VANDERBILT,	*CONNECTICUT,*
CAPT. DEMING.	CAPT. SENISKY.

FARE FIFTY CENTS,

Whole State Rooms $1.50, Half Rooms 75 Cents.

Leave Daily, Saturdays excepted,

FROM PIER 44 NORTH RIVER,

Bet. Charlton and Spring Streets, at 6 P. M.

The Steamers will leave as above, connecting at Albany and Troy with Albany and Susquehanna, New York Central, Renssaelaer and Saratoga, and Troy and Boston Railroads.

Through Tickets and Baggage Checked to all Points.

GENERAL OFFICE, PIER 44 N. R.

C. D. HANCOX, *Gen'l Agent.*

SHORT-TRIP GUIDE — ANNOUNCEMENTS.

TRAVELERS' OFFICIAL GUIDE

OF THE

RAILWAYS

AND

STEAM NAVIGATION LINES

IN THE

UNITED STATES & CANADA.

The only Guide recognized by the United States Government, and railroad officers, as the standard authority for time, distances, and other statistics relative to railroads.

PUBLISHED MONTHLY,

under the auspices of the *General Ticket Agents' Association*, and sold by all news agents and periodical dealers throughout the United States and Canada; also, at the various railroad depots, and on trains. This is the

BEST ADVERTISING MEDIUM

in the country, and from its large circulation, affords unusual advantages to merchants and manufacturers.

For further particulars, as to prices of advertising, etc., application should be made to the

GENERAL EDITOR,

237 & 239 Dock Street,
Philadelphia, Pa.

SHORT-TRIP GUIDE—ANNOUNCEMENTS

FIFTH AVENUE THEATRE,

(TWENTY-FOURTH STREET,)

ONE DOOR FROM BROADWAY,

Near the Fifth Avenue, St. James, Hoffman, Grand, and other Fashionable Hotels,

NEW YORK CITY.

MR. *AUGUSTIN DALY.* - - PROPRIETOR.

Has, since its opening, produced a constant succession of

DRAMATIC SURPRISES:

SOCIETY PLAYS,

CONTEMPORANEOUS COMEDIES,

PARISIAN SENSATIONS, &c.

Perfect in every detail of presentation. and always interpreted by

THE LEADING COMEDY COMPANY IN AMERICA.

☞ Among the most pronounced successes of the management, may be mentioned some of world-wide reputation:

"FROU-FROU," "FERNANDE," "SARATOGA,"

And the favorite of a whole year,

"DIVORCE."

☞ This theatre is unsurpassed in the fashionable quality of its attendance, as in the novelty and brilliancy of its performances.

The *New York Herald* of Dec. 17th says:—" Were a stranger of refinement coming to New York to ask us where he would probably feel most at home for amusement, and happiest for an evening, without great exertion or intense excitement, we should unhesitatingly and sans-invidiousness, say--at the FIFTH AVENUE THEATRE."

SHORT-TRIP GUIDE.—ANNOUNCEMENTS.

OLYMPIC THEATRE,

622 & 624 Broadway,

NEW YORK CITY.

Lessee and Manager, JAMES E. HAYES.

THE HOME OF PANTOMIME,
CULTURED DRAMA, AND
CHASTE SPECTACLE.

The scene of the early triumphs of

Mrs. JOHN WOOD, JOSEPH JEFFERSON,
E. A. SOTHERN,
DION BOUCICAULT and AGNES ROBERTSON.

Added to this list of the Olympic's Phenomenal Luminaries,
is

Mr. G. L. FOX,

At present the reigning American Star, in his famous

PANTOMIME,

"HUMPTY DUMPTY,"

Which has been performed at this Theatre over 900 times, and is still running.

This Theatre is patronized by the best society of New York, the Company is selected from the best European material, and the entertainment is invariably of the brilliancy and refinement demanded by educated and fashionable tastes.

SHORT-TRIP GUIDE.—ANNOUNCEMENTS.

STEVENS & CO.,
Jewelers & Silversmiths
859 BROADWAY,

(Second door above Seventeenth Street.)

NEW YORK.

☞ Specialties of RARE DIAMONDS AND OTHER FINE JEWELRY.

GOLD AND SILVER WORK

Of the most artistic patterns and elaborate finish, and *Patronage of the First Taste and Fashion.*

SHORT-TRIP GUIDE.—ANNOUNCEMENTS.

NASH & FULLER,

DINING, LUNCH, OYSTER,

AND

COFFEE ROOMS,

39, 40 & 41 PARK ROW,

AND

147, 149 & 151 NASSAU STREET,

(TIMES BUILDING,)

NEW YORK.

Largest Restaurant in the United States.

EXTENSIVE AND CONVENIENT

BILLIARD SALOONS CONNECTED,

In which is generally to be seen some of the finest AMATEUR AND PROFESSIONAL PLAYING known to the world of amusement.

THE GILSEY HOUSE,

Cor. Broadway and 29th St.,

NEW YORK.

One of the most complete and thoroughly-appointed Hotels in America, with *Passenger Lift* and all modern conveniences and improvements; and most conveniently located for all purposes of residence or visit.

BRESLIN, GARDNER & CO.,
Proprietors.

In connection with the Gilsey House, the

GRAND UNION HOTEL,

SARATOGA SPRINGS,

NEW YORK.

One of the largest in the world, situated in grounds of peculiar extent and beauty, containing

700 Private Parlors and Bed-Rooms,

and second to none on either continent, in every detail of luxurious comfort.

BRESLIN, GARDNER & CO.

BREVOORT HOUSE,

NEW YORK.

This well known Hotel is located on Fifth Avenue, corner of Eighth Street, near Washington Park, one of the most delightful locations, combining the quiet retirement of a private mansion, with easy access to all parts of the City. The Brevoort has always been a favorite with Europeans visiting the United States; the plan upon which it is kept, being such as to specially commend it to those accustomed to European habits.

CLARK & WAITE,
Proprietors.

SHORT-TRIP GUIDE.—ANNOUNCEMENTS.

THE LAUREL HOUSE,

KAUTERSKILL FALLS,

(1½ miles west of Mountain House), CATSKILL MOUNTAINS.

J. L. SCHUTT, Proprietor.

Possesses great beauty of location, with unequaled opportunities for examining the Falls; secluded walks and delightful retreats; the Clove and other matchless drives. Good Trout Fishing in the neighborhood. Recently enlarged, and in perfect order. Carriages and an Agent in attendance at the Cars and Boats, at Catskill.

POWELL HOUSE,

HOTEL AND POSTING HOUSE.

Point of Departure of all Conveyances for the Mountains,

CATSKILL LANDING,

GREENE CO., NEW YORK.

JOHN T. HUNTLEY, . . Proprietor.

ROSSIN HOUSE HOTEL,

TORONTO, CANADA.

This Splendid Commodious Hotel (opened by the undersigned on the 1st of August, 1867,) is finished and furnished with every regard to comfort and luxury; has hot and cold water, with Baths and Closets on each floor. The Parlors and Bed Rooms are large and well ventilated, and arranged for private parties and families. The aim has been to make this the most unexceptionable first-class Hotel in Canada.

The undersigned trusts that his long experience in the CLIFTON HOUSE at Niagara Falls, will give confidence to his friends and the traveling public that they will receive every attention and comfort, with reasonable charges, at this new and elegant House.

G. P. SHEARS.

CONGRESS HALL,

CAPE MAY, NEW JERSEY.

J. F. CAKE, Proprietor.

This first-class and well-known Hotel—always one of the most popular at this great sea-side resort, offers increased attractions for the season of 1872.

Recent additions give this House the unequalled advantage of

A SEA FRONT OF 900 FEET.

BRADY'S
National Photographic Portrait
GALLERIES,

BROADWAY & 10th STREET,

NEW YORK,

627 PENNSYLVANIA AVENUE,

WASHINGTON, D. C.

☞ Largest and most perfect collection of Portraits of American Celebrities, in Literature Statesmanship, the Arts, Army and Navy, and all other departments, to be found in the United States; embracing an aggregate of 2,000 valuable portraits; and expected to be adopted by the nation, at an early day, as material for an American Pantheon.

Admission free, and all courtesies extended.

PORTRAITS WITH SHORT DELAY AND IN THE HIGHEST STYLES OF THE ART.

"THE GALAXY is about as near perfection as anything can be."—*Daily Register, New Haven, Conn.*

THE GALAXY
IS THE
BEST AMERICAN MAGAZINE.

☞ NO FAMILY CAN AFFORD TO DO WITHOUT IT.

It gives more Good and Attractive Reading Matter for the money than any other Periodical or Book published in the country.

The leading newspapers pronounce THE GALAXY the Best and most Ably Edited American Magazine.

THE GALAXY meets the wants of every member of the Family.

It contains Thoughtful Articles by our ablest writers.
It contains Sketches of Life and Adventure.
It has Serial Stories by our best Novelists.
It has Short Stories in each number.
It has Humorous Articles in each number which are a constant source of delight to the public.
In each number is a complete Review of the World of Literature.
The New Department of "*Scientific Miscellany*" appears in each number.

WHAT THE LEADING PAPERS SAY.

"First of all in attractions we place THE GALAXY: it has succeeded better than any of its rivals."—*Standard, Chicago.*

"Well sustains its reputation for vigorous and racy writing."—*New York Tribune.*

"THE GALAXY is always more a magazine than any other."—*Independent.*

"A model periodical; a credit to American periodical literature."—*Philadelphia Press.*

"We are inclined to believe that more downright good literature is crowded between the covers of THE GALAXY than any other American magazine can boast of."—*Chicago Times.*

"There is not a dull page between its covers."—*New York Times.*

"The variety of its contents, their solid worth, their brilliance, and their great interest, make up a general character of great excellence for every number."—*Post, Boston.*

"We are glad to be able to congratulate THE GALAXY on surpassing even the high standard with which, as one of our leading magazines, we credit it."—*Evening Mail, New York.*

"Always ably edited, and remarkable for the good judgment displayed in the selection of current topics for discussion. In this way it quite eclipses the more conservative periodicals of the day."—*Boston Journal, Mass.*

"The reported increase in the circulation of THE GALAXY is hardly to be wondered at, for it is certainly the best of American magazines."—*Express, Buffalo, N. Y.*

NOW IS THE TIME TO SUBSCRIBE.

Price 35 cts. per Number. Subscription Price, $4 per year.

Address,
SHELDON & COMPANY, 677 Broadway, New York.

GRANT LOCOMOTIVE WORKS.

(D. B. GRANT, PRESIDENT.)

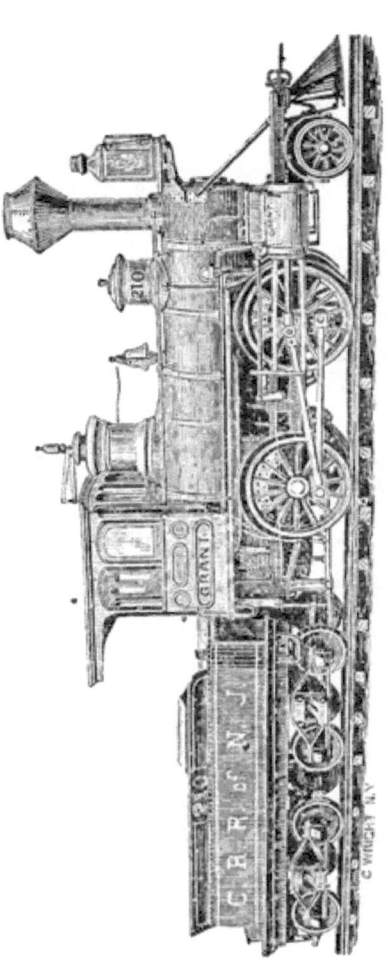

PATERSON, NEW JERSEY.

New York City Office, 33 Wall Street.

Receivers of the great Gold Medal of the first-class, at the Paris Exposition of 1867, for perfection of Locomotive manufacture.

www.ingramcontent.com/pod-product-compliance
Lightning Source LLC
Chambersburg PA
CBHW031422230426
43668CB00007B/396